Injection Molding

Rong Zheng · Roger I. Tanner
Xi-Jun Fan

Injection Molding

Integration of Theory and Modeling Methods

Springer

Rong Zheng
School of Aerospace,
Mechanical and Mechatronic Engineering
The University of Sydney
Sydney
Australia
e-mail: rong.zheng@uni.sydney.edu.au

Xi-Jun Fan
School of Aerospace,
Mechanical and Mechatronic Engineering
The University of Sydney
Sydney
Australia
e-mail: xijun.fan@sydney.edu.au

Roger I. Tanner
School of Aerospace,
Mechanical and Mechatronic Engineering
The University of Sydney
Sydney
Australia
e-mail: rit@aeromech.usyd.edu.au

ISBN 978-3-642-21262-8 e-ISBN 978-3-642-21263-5
DOI 10.1007/978-3-642-21263-5
Springer Heidelberg Dordrecht London New York

Cover design: eStudio Calamar, Spain

Printed on acid-free paper

Springer is part of Springer Science+Business Media (www.springer.com)

Preface

The purpose of this book is to introduce the reader to some topics relevant to the modeling of the injection molding process. Injection molding processing links to various scientific and engineering disciplines such as rheology, mechanical and chemical engineering, polymer science and computational methods. While it is impossible to address every aspect of such a broad range of issues, this book is intended to address the up-to-date status of fundamental understanding and simulation technologies, without losing sight of still useful classical approaches.

The book is organized as follows. In Chap. 1, a general overview of the injection molding stages, material classification, rheological characterization of polymers, and the development of numerical simulation methods is presented as an introduction. In Chap. 2, fundamentals of rheology are described, focused on constitutive models useful in practice. Chapter 3 describes assumptions and mathematical models for filling and packing phases of the injection molding process. Chapter 4 is devoted to flow-induced crystallization and the processing–morphology–properties relationship. This is a very active field, which has attracted great attention in recent years. Chapter 5 discusses the modeling of flow-induced orientation distribution in injection molded fiber-reinforced polymers, highlighting the efforts to simulate particle–particle interactions. The application of the multilevel micro–macro approach to fiber suspensions is also demonstrated. Chapter 6 presents methods to predict shrinkage and warpage. Micromechanics for mechanical property predictions is also included in this chapter. Finally, Chap. 7 deals with the application of boundary integral equations to the mold cooling analysis. Chapter 8 provides a guide to some computational techniques used in simulations of melt flow in mold filling and packing, deformation of the solid parts, and mold cooling. Cartesian tensors are used in this book for conciseness.

The authors wish to acknowledge with deep appreciation our many colleagues and research collaborators: Nhan Phan-Thien, Peter Kennedy, Graham Edward, Pengwei Zhu, Ahmad Jabbarzadeh, Chitiur Hadinata, Shaocong Dai, Huashu Dou and Duane Lee Wo, who have contributed to the research outcomes which the book is based on. We also wish to thank Charles Tucker, Jay Schieber, Gilles Regnier, Rene Fulchiron, Didier Delaunay, Vito Leo, Gerhard Eder, Gerrit Peters,

Zhilaing Fan, Huagang Yu, Clinton Kietzman, Xiaoshi Jin, Zhongshuang Yuan, Franco Costa and Edwin Klompen for many fruitful interactions and discussions. Thanks also due to the Australian Research Council (ARC), the Cooperative Research Centre for Polymers (CRC-P) and Moldflow Pty Ltd (now, Autodesk Australia) for the financial support to the authors' research during the past years.

Finally, our personal thanks are extended to our families for the love, support and encouragement.

June 2011

R. Zheng
R. I. Tanner
X.-J. Fan

Acknowledgments

We acknowledge with gratitude the assistance of the following organizations and individuals.

American Chemical Society

Figure 4.10 is reproduced with permission from the article by A. Jabbarzadeh and R.I. Tanner, Macromol 43: 8136–8142. Copyright 2010.

American Institute of Physics for the Society of Rheology

Figure 4.3 is reproduced with permission from the article by R. Zheng and P.K. Kennedy, J Rheol 48: 823–842. Copyright 2004.

Elsevier Science Publishers B.V.

Experimental data in Fig. 4.4 is reproduced with permission from the article by R. Mendoza, G. Régnier, W. Seiler, and J.L. Lebrun, J Mater Sci 30: 5002–5012. Copyright 2003.

Figures 5.2, 6.4 and 6.6 are reproduced with permission from the article by R. Zheng, P.K. Kennedy, N. Phan-Thien and X.-J. Fan, J Non-Newtonian Fluid Mech 84: 159–190. Copyright 2009.

Figure 5.3 is reproduced with permission from the article by N. Phan-Thien, X.-J. Fan, R.I. Tanner and R. Zheng, J Non-Newtonian Fluid Mech 103: 251–260. Copyright 2002.

John Wiley and Sons Inc.

Figure 4.5 is reproduced with permission from the article by J.-F. Luyé, G. Regnier, P.H. Le Bot, D. Delaunay and R. Fulchiron, J Appl Polym Sci 79: 302–311. Copyright 2001.

Figure 8.6 is reproduced with permission from the article by F. Dupret and L. Vanderschuren, AICHE J 34: 1959–1972. Copyright 1988.

Nova Science Publishers Inc.

Figure 8.11 is reproduced with permission from the article by X.-J. Fan, R.I. Tanner and R. Zheng, in P.H. Kauffer (ed.) Injection Molding: Process, Design, and Applications. Copyright 2011.

Sage Science Press

Figures 6.7 and 7.1 are reproduced with permission from the article by R. Zheng, N. McCafferey, K. Winch, H. Yu, P.K. Kennedy, J Thermoplastic Compos Mater 9: 90–106. Copyright 1996.

Society of Plastics Engineers Inc.

Figure 1.2 is reproduced with permission from the article by P.K. Kennedy and R. Zheng ANTEC'2002, 48: 1–7. Copyright 2002.

Figures 5.4 and 5.5 are reproduced with permission from the article by R. Zheng, P.K. Kennedy, X.J. Fan, N. Phan-Thien, R.I. Tanner, ANTEC'2000, 46: 671–674. Copyright 2000.

Figure 6.9 is reproduced with permission from the article by A. Bakharev, R. Zheng, Z. Fan, F.S. Costa, X. Jin, P.K. Kennedy, ANTEC' 2005, 51: 511–515. Copyright 2005.

Springer Science+Business Media B.V.

Figures 4.1, 4.2, 4.7, 4.8 and 4.9 are reproduced with permission from the article by R. Zheng, R.I. Tanner, D. Lee Wo, X.-J. Fan, C. Hadinata, F.S. Costa, P.K. Kennedy, P. Zhu and G. Edward, Korea-Australia Rheol J 22: 151–162. Copyright 2011.

Thanks are also due to Duane Lee Wo for supplying experimental data which we have used in Fig. 4.6, Peter Kennedy for Figs. 8.1 and 8.6, Y.-Q. Zheng for help with drawing Fig. 1.1, and Zhiliang Fan for commenting on Sect. 8.33.

Contents

Chapter 1
Introduction

1.1 Injection Molding

Injection molding, defined as a cyclic process for producing identical articles from a mold, is the most widely used polymer processing operation. The main advantage of this process is the capacity of repetitively fabricating parts having complex geometries at high production rates.

1.1.1 Injection Molding Machine

Injection molding machine history goes back to the "packing machine" invented by the Hyatt brothers, who received a US patent in 1872 for the invention of a machine to mold camphor-plasticized cellulose nitrate (Rubin 1972). The machine contained a basic plunger to inject the plastic into a mold through a heated cylinder. The plunger-type device was replaced by a screw injection machine in 1946. Since then, several improvements have led to the reciprocating screw injection molding machines commonly used today. Figure 1.1 shows a typical reciprocating screw injection molding machine. Primarily it consists of two distinct units: an injection unit comprising a hopper, a rotating screw and a heated barrel, and a clamping unit containing the mold that is typically made of two halves. More details on injection molding machines can be found in Johannaber (1994).

1.1.2 Injection Molding Cycle

In operation, plastic granules are fed to the machine through the hopper. Upon entrance into the barrel, the screw rotates and moves the granules forward in the screw channels. The granules are forced against the wall of the barrel, and melt

R. Zheng et al., *Injection Molding*, DOI: 10.1007/978-3-642-21263-5_1,

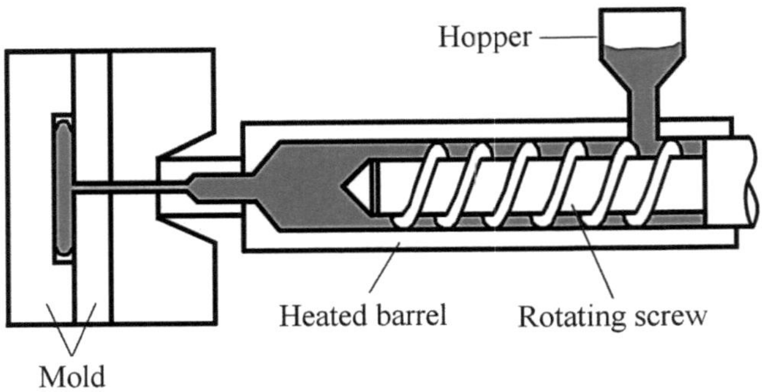

Fig. 1.1 A sketch of a reciprocating screw injection molding machine

due to both the friction heat generated by the rotating screw and the conduction from the heating units along the barrel. The molten material is conveyed to the tip of the screw. During this time, pressure develops against the "closed-off" nozzle and the screw moves backward to accumulate a reservoir of melt at the front end of the screw barrel. When the desired volume of the melt is obtained, the screw rotation stops, signifying the end of the stage for production of the flow of molten polymer. This stage of process is also called the plasticizing stage. Then the injection stage begins. The injection stage is characterized by the following four phases.

1.1.2.1 Filling

The clamp unit keeps the empty mold closed. The screw moves forward as a ram and forces the melt into the mold cavity.

1.1.2.2 Packing

When the mould is filled, the screw is held in the forward position or moves with a small displacement to maintain a holding pressure, during which time the material cools down and shrinks, allowing a little more material to enter the mold to compensate for volumetric shrinkage of the material. The packing phase is also called the *holding* phase.

1.1.2.3 Cooling

At some stage of packing, the gates have completely frozen. Eventually the cavity pressure is reduced to zero or a low value. The part continues to cool down and to solidify. Meanwhile, the screw starts rotating and moving backward, and the next plasticizing stage takes place.

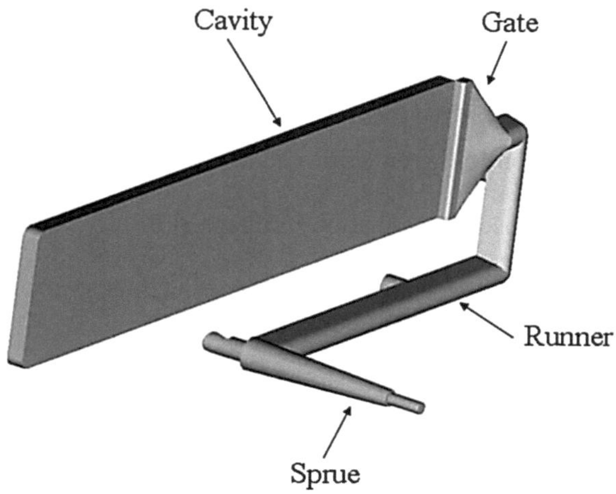

Fig. 1.2 Components of an injection molded part (Reproduced from Kennedy and Zheng 2002, with permission from Society of Plastics Engineers Inc.)

1.1.2.4 Ejection (demolding)

When sufficient cooling time has been allowed for the part to solidify and to become stiff enough, the mould opens and the part is ejected.

The mold then closes and the injection cycle starts again.

1.1.3 Feed System

The polymer melt injected from the nozzle of the injection molding machine flows through the sprue, runner and gate, and finally enters the cavity. The *feed system* is a generic term for the sprue, runner and gate. The feed system is usually included in the flow analysis as part of the flow channel upstream to the cavity (Fig. 1.2).

Runners can be classified into two types, namely the *cold runner* and *the hot runner*. The mold temperature for the cold runner is similar to the mold temperature for the cavity. The material in the cold runner will eventually solidify, and be ejected with each cycle. The hot runner maintains the polymer at a melt temperature, either by insulation or by heating. This keeps the material fluid. Hot runners are not ejected with each cycle.

1.1.4 The Need for Rheological Information

Two key problems of injection molding processing are (i) shaping such parts that meet the desired dimensional tolerances and surface finish, and (ii) making fabricated parts with desirable mechanical, thermal, optical and other ultimate properties. The dimensional stability, surface finish and ultimate properties are the desired qualities of the injection molded products. To investigate these problems requires understanding of the following relations and effects:

1. relations between deformation, local temperature and applied stress within the material;
2. influence of the thermo-mechanical history experienced by the material on its internal structure;
3. Impact of the processing-induced internal structure on the flow behavior of the final properties of the material.

Research of these relations and effects is the province of rheology.

1.2 Classification of Polymers

Polymer chains may be linear, or possibly with additional chemical groups forming branches along the primary chain. If a polymer is made up of the same repeating units of monomers, it is called a homopolymer; otherwise, if it is made up of different types of monomers arranged in some sequence, then it is called a copolymer.

The number of the monomer units in a polymer chain can be very large (up to 10^5). The molecular structure leads to two types of materials: thermoplastics and thermosets. Thermoplastics turn to a liquid when heated, and they solidify when cooled sufficiently. Their molecules are not chemically joined with each other during processing. The materials can be reversibly softened and hardened by heating and cooling. Thermosets are polymers that chemically react during processing to form a three-dimensional cross-linked polymer chain network. The chemical reaction is irreversible. Once hardened, the material cannot be converted back to a melt by heating.

Both thermoplastics and thermosets are used in injection molding. The main difference in processing is the mold temperature. For thermoplastics, the mold walls are colder than the melt, while for thermosets the mold walls are hotter than the material in the cavity.

This book will only consider thermoplastics. Most of the mathematical simulation methods described later are, however, applicable to thermosets as well. The reader can see Sidi and Kamal (1982) for some discussions on injection molding of thermosets.

Based on their chain conformation and morphology, thermoplastic polymers can be classified as

- Amorphous,
- Semi-crystalline,
- Liquid Crystalline.

Amorphous polymers comprise randomly configured molecular chains. Upon cooling, they change from rubber-like materials to glassy materials. The temperature at which the transition occurs is called the glass transition temperature (T_g), At this temperature, there is a change in the slope of the specific volume/temperature curve.

Under some favorable circumstances, some polymers can align their molecules in a regular lattice shape. Such materials are said to be crystallizable. Often, if the cooling starts from the melt phase, the crystallization is only partial, and hence the materials possess both crystalline and amorphous structure and so are called semi-crystalline. There is a jump of specific volume at a transition zone in the specific volume/temperature curve below the melting temperature (T_m), and then there is a region of tough solid. The mechanism of the crystallization is still the object of intense discussions (Tanner and Qi 2005; Pantani et al. 2005).

Liquid crystalline polymers (LCPs) exhibit ordered molecular arrangements in the melt state. The polymer molecules of LCPs are generally semi-rigid or contain rigid units. The rigid units are either along the backbone or in the side branches. LCPs are viscoelastic and anisotropic. The interested reader can learn details about LCPs from Feng and Leal (1997), Larson (1999), Fan et al. (2003), and Klein et al. (2008).

Many materials used in injection molding processing are not neat resins, but composite materials. The term "composite material" refers to a structure made up of two or more components, insoluble in one another, which, when combined, enhance the behavior of the resulting material. Composite properties are dominated by the microstructure of the fabricated part rather than the properties of the constituent materials. Among various possible reinforcements, glass and carbon fibers are commonly used in injection molding. Reinforcing fibers could be roughly divided into short fibers and long fibers. Short fibers are slender, but their typical length is smaller than the typical dimension of the apparatus. The aspect ratios (length/diameter) of short-fibers are smaller than 100, typically about 20, while the aspect ratios of long fibers are greater than 100. In general, the mechanical properties of a short-fiber composite are not as good as a composite with long fibers, but short-fiber composites are easier to be processed at high production rate with methods such as injection molding. We shall only consider the short fibers in this book.

1.3 Rheological Characterization of Polymer Fluids

1.3.1 Subject and Goals

Rheology is generally understood to be the study of deformation and flow of mater, especially of non-classical materials. In its origin, the term "rheology" was inspired by Heraclitus's quote "everything flows", since the main root of the word

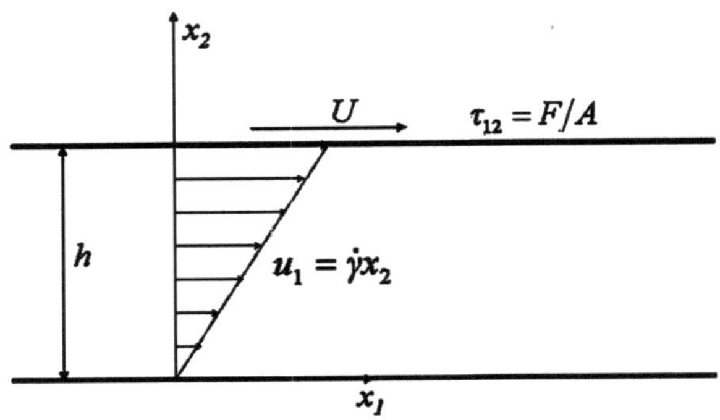

Fig. 1.3 Schematic of simple shear flow

"rheo" in Greek means "to flow" (Reiner 1964). Historical reviews of this research area can be found in Tanner and Walters (1998) and Denn (2004).

The central problem of rheology is to establish the relationship between applied forces and geometrical effects induced by these forces at a point. The mathematical description of the relationship is called the *constitutive equation*, or the *rheological equation of state*. It is a model of physical reality. The recent subject of rheology has tended to establish relationships between macroscopic rheological properties of material and its micro-level structure (Tanner 2009).

Fundamental rheological theories provide a basis for predictions of the mechanical and thermal behavior of materials in processing. Rheological measurements provide us with properties of materials that can be used in numerical simulations to solve applied problems. Rheology is also a structural characterization method. It gives physical meaningful parameters that can be correlated with the structure of material and provide insight into the processing–structure–property relations.

1.3.2 Some Rheological Phenomena

1.3.2.1 Shear-Rate-Dependent Viscosity

The most important property of a fluid material for engineering calculations is the fluid viscosity,η defined as the ratio of the shear stress τ_{12} to the shear rate $\dot{\gamma}$, i.e.,

$$\eta = \tau_{12}/\dot{\gamma}. \tag{1.1}$$

As show in Fig. 1.3, where the fluid is contained between two parallel plates separated by a distance *h*; the shear flow is generated by sliding the top plate with a constant velocity U in x_1 direction. In this case, the shear rate $\dot{\gamma} = U/h$, and the shear stress equals the tangential force on the top plate (*F*) divided by the fluid contact area (*A*).

Fluids for which shear stress is directly proportional to shear rate are called *Newtonian fluids*; otherwise, they are *non-Newtonian Fluids.* The Newtonian fluid is also called the Navier–Stokes fluid, and the non-Newtonian fluid can be defined as any fluid whose behavior is not characterized by the Navier–Stokes equations (Boger and Walters 1993). The viscosity of a Newtonian fluid is a constant depending mainly on the temperature and, less markedly, the pressure. Most fluids with long chain microstructures such as molten polymers are non-Newtonian. For most molten polymers, the viscosity is a decreasing function of the shear rate, known as shear thinning. The opposite of the shear thinning is the shear thickening, which has also been observed in some highly concentrated suspensions, due to the formation of clusters of the suspending particles. While both the shear-thinning and shear thickening fluids usually behave like the Newtonian model at low shear rate region, some particle-filled thermoplastics and gel materials appear to show an infinite value of viscosity as the shear rate approaches zero. For such fluids, no detectable flow takes place until the local stresses exceed a critical value. The critical value is called a *yield stress*.

1.3.2.2 Normal Stress Differences

Polymers usually show unequal normal stress components in steady shear flow (for a Newtonian fluid in steady shear flow, the three normal stress components are always equal). With the three normal stress components σ_{11}, σ_{22} and σ_{33}, we can define two normal stress differences:

$$N_1 = \sigma_{11} - \sigma_{22}, \tag{1.2}$$

and

$$N_2 = \sigma_{22} - \sigma_{33}, \tag{1.3}$$

where N_1 and N_2 are the *first* and the *second normal stress differences*, respectively.

The inequality of normal stresses is responsible for a number of visually noticeable Non-Newtonian effects. These include the Weissenberg rod-climbing effect where the fluid climbs up a rotating rod rather than dipping near the rod, and the extrudate swell, see Tanner (2000) and Boger and Walters (1993) for more details.

1.3.2.3 Viscoelasticity

In steady shearing, the materials can appear to be viscous and inelastic. In many cases (such as in elongational flows) the materials are often intermediate between an "elastic solid" and a "viscous liquid". We call the material a *viscoelastic fluid*. A viscoelastic fluid has a characteristic time scale λ. It is customary to define a "Deborah number" (Reiner 1964), *De*, as the ratio of a characteristic time of the fluid to a characteristic time of the flow process t_{flow}, i.e.,

$$De = \lambda / t_{\text{flow}}. \tag{1.4}$$

The Deborah number represents the transient nature of the flow relative to the material time scale. In fast processes (t_{flow} is very small compared to the fluid's characteristic time and hence $De \gg 1$), the material will response like an elastic solid, while in slow processes ($De \ll 1$), a liquid-like response is expected.

Another important dimensionless group, the Weissenberg number (White 1964), is defined as the product of the characteristic time of the fluid and the characteristic shear rate of the flow, i.e.,

$$Wi = \lambda\dot{\gamma}. \tag{1.5}$$

The Weissenberg number compares the elastic forces to the viscous effects. It is usually used in steady flows. One can have a flow with a small *Wi* number and a large *De* number, and vice versa. Sometimes the characteristic time of the flow in the definition of the Deborah number has been taken to be the reciprocal of a characteristic shear rate of the flow; in these cases, the Deborah number and the Weissenberg number have the same definition. Pipkin's diagram (see Fig. 3.9 in Tanner 2000) classifies shearing flow behavior in terms of *De* and *Wi*, and provides a useful guide for the choice of constitutive equations.

Viscoelastic phenomena may be described through three aspects, namely *stress relaxation, creep* and *recovery*. Stress relaxation is the decline in stress with time in response to a constant applied strain, at a constant temperature. Creep is the increase in strain with time in response to a constant applied stress, at a constant temperature. Recovery is the tendency of the material to return partially to its previous state upon removal of an applied load. The material is said to have *memory* as if it remembers where it came from. Because of the memory effect, in transient flows the behavior of viscoelastic fluids will be dramatically different from that of Newtonian fluids. Viscoelastic fluids are full of instabilities. Some examples include instabilities in Taylor-Couette flow, in cone-and-plate and plate-and-plate flows (Larson 1992). The extrudate distortion, commonly called melt fracture, is a notorious example of viscoelastic instability in polymer processing. The viscoelastic instability in injection molding can result in specific surface defects such as "tiger stripes" (Bogaerds et al. 2004).

1.3.3 Solidification Rheology

Injection molding processing involves both a molten flow phase and a solidification phase. The major challenges of constitutive modeling of the liquid–solid transition involve two related topics. First, one needs to consider how the flow and thermal history influence the structure of the fluid. Secondly, one needs to understand how the changes in the internal structure stiffen the material. There are some investigations on the topics for semicrystalline materials and a variety of approaches or models have been proposed by different authors. Most of these results are reviewed by Tanner and Qi (2005) and Pantani et al. (2005). However, there is still a lack of a complete and fundamental theory to describe the stiffening effect of crystallization.

1.4 Development of Numerical Simulations for Injection Molding

Some major features of injection molding processes are:

1. geometric complexity of the flow channel;
2. complex properties of the material;
3. transient flow with an advancing flow-front;
4. non-Newtonian fluid behavior;
5. non-isothermal conditions;
6. effects of high pressure and compressibility;
7. liquid/solid phase change during processing.

Given the above complications, one is forced to use numerical methods for the solution of the injection molding problems. Early numerical simulations for mold flow began with the work of Toor et al. (1960). Pearson (1966) outlined a finite difference scheme for solving a problem of a geometrically simple center-gated injection-molded part. This work, together with other pioneering studies by Harry and Parrott (1970), Kamal and Kenig (1972), Berger and Gogos (1973), Lord and Williams (1975), Williams and Lord (1975), Pearson and Richardson (1983) and Agassant et al. (1991) established mechanical principles and computational models for polymer injection molding processing. Science-based commercial software for injection molding simulation emerged due to the efforts of Austin (Austin 1987; Whiteford 1992), Hieber and Shen (1980), Wang et al. (1986), Kamal (1987), (Nakano 1998; Nakano 2000), Chang and Yang (2001) and their coworkers. The history of development of commercially available numerical technology has been reviewed in Kennedy (2008).

Despite the rapid progress in injection molding simulation, the solution of such problems remains challenging. In particular, the understanding of modeling of processing–morphology–property relationships, the efficient and accurate prediction of shrinkage, orientation, stresses and warpage, and the treatment of geometric complexity are still active research areas. For an extensive survey of the state of the science and technology for the injection molding process, see the recent book edited by Kamal et al. (2009).

Chapter 2
Fundamentals of Rheology

2.1 Introduction to Basic Concepts

The term "rheology" dates back to 1929 (Tanner and Walters 1998) and is used to describe the mechanical response of materials. Polymeric materials generally show a more complex response than classical Newtonian fluids or linear viscoelastic bodies. Nevertheless, the kinematics and the conservation laws are the same for all bodies. The presentation here is condensed; one may consult other books for amplification (Bird et al. 1987a; Huilgol and Phan-Thien 1997; Tanner 2000). We begin with kinematics.

2.1.1 Kinematics

There are two viewpoints that one may take to describe the fluid motion: the *Lagrangian* and the *Eulerian viewpoints.*

In the Lagrangian viewpoint, one keeps track of a fluid particle, which, at time t, is located at a position x_i ($i = 1, 2, 3$) in Cartesian coordinates. The trajectory of the particle would be given by x_i (t), where x_i is a dependent variable and t is an independent variable. The velocity of the fluid particle is therefore given by

$$u_i = \frac{dx_i(t)}{\mathrm{d}t}. \tag{2.1}$$

In the Eulerian viewpoint, the quantities in the flow field are described as a function of fixed position x_i and time t. Specifically, the flow velocity is described as

$$u_i = u_i(x_i, t), \tag{2.2}$$

where x_i is an independent spatial variable along with time t. In Lagrangian coordinates the acceleration is

R. Zheng et al., *Injection Molding*, DOI: 10.1007/978-3-642-21263-5_2,

$$a_i = \frac{\mathrm{d}u_i(t)}{\mathrm{d}t} = \frac{\mathrm{d}^2 x_i(t)}{\mathrm{d}t^2}. \tag{2.3}$$

The Eulerian acceleration field is given by

$$a_i = \frac{\partial u_i}{\partial t} + u_j \frac{\partial u_i}{\partial x_j} \equiv \frac{\mathrm{D}u_i}{\mathrm{D}t}, \tag{2.4}$$

where the operator $\mathrm{D}/\mathrm{D}t = \partial/\partial t + u_j \partial/\partial x_j$ is called the material derivative. Here and elsewhere, repeated indices imply summation unless stated otherwise. The rate of change of the velocity component u_i with respect to the coordinate x_j, $\partial u_i/\partial x_j$, is called the velocity gradient, written as

$$L_{ij} = \frac{\partial u_i}{\partial x_j}, \tag{2.5}$$

or

$$\mathbf{L} = (\nabla \mathbf{u})^T. \tag{2.6}$$

One can split $\mathbf{L}$ into symmetric and anti-symmetric parts

$$L_{ij} = D_{ij} + W_{ij}, \tag{2.7}$$

where

$$\begin{aligned} D_{ij} &= \frac{1}{2}\left(\frac{\partial u_i}{\partial x_j} + \frac{\partial u_j}{\partial x_i}\right) = \frac{1}{2}(L_{ij} + L_{ji}), \\ W_{ij} &= \frac{1}{2}\left(\frac{\partial u_i}{\partial x_j} - \frac{\partial u_j}{\partial x_i}\right) = \frac{1}{2}(L_{ij} - L_{ji}). \end{aligned} \tag{2.8}$$

The symmetric part D_{ij} is called the rate of deformation tensor, and the anti-symmetric part W_{ij} is called the vorticity tensor. The rate of deformation tensor describes the rate at which neighboring particles move relative to each other, irrelevant to superposed rigid rotation. The vorticity is a measure of the local rate of rigid rotation.

The invariants of the rate of deformation tensor are:

$$I_D = D_{ii}, \tag{2.9}$$

$$II_D = D_{ij}D_{ij}, \tag{2.10}$$

$$III_D = D_{ij}D_{jk}D_{ki}. \tag{2.11}$$

The first invariant I_D represents rate of change of volume, which is zero for incompressible fluids. The third invariant III_D is zero for plane flows. The second invariant II_D represents a mean rate of deformation including all shearing and extensional components. It is convenient to define, for all flows, a *generalized strain rate* as

$$\dot{\gamma} = (2II_D)^{1/2} = \left(2D_{ij}D_{ij}\right)^{1/2} \text{ or } \dot{\gamma} = \left(2tr(\mathbf{D}^2)\right)^{1/2}. \tag{2.12}$$

Given an initial (reference) position $\mathbf{X}$ of a particle, which is subsequently at an Eulerian location $\mathbf{x}$, the gradient of $\mathbf{x}$ with respect to $\mathbf{X}$, written as $\mathbf{F} = (\partial\mathbf{x}/\partial\mathbf{X})^T$ $(F_{ij} = \partial x_i/\partial X_j)$, is called the deformation gradient tensor, and the tensor

$$C_{ij} = F_{ki}F_{kj} = \frac{\partial x_k}{\partial X_i}\frac{\partial x_k}{\partial X_j} \tag{2.13}$$

is called the Cauchy-Green strain tensor, which is a measure of the strain that the fluid experiences. A *generalized strain* is then defined as

$$\gamma = (C_{kk} - 3)^{1/2} \text{ or } \gamma = (\text{tr}C - 3)^{1/2}. \tag{2.14}$$

2.1.2 Viscometric and Extensional Flows

Consider a steady simple shear flow with the kinematics given by $u_1 = \dot{\gamma}_{13}x_3$, $u_2 = 0$, $u_3 = 0$ where the shear rate $\dot{\gamma}_{13}$ is a constant. This flow has the rate of deformation

$$\mathbf{D} = \begin{bmatrix} 0 & 0 & \frac{1}{2}\dot{\gamma}_{13} \\ 0 & 0 & 0 \\ \frac{1}{2}\dot{\gamma}_{13} & 0 & 0 \end{bmatrix}. \tag{2.15}$$

The generalized strain rate is $\dot{\gamma} = |\dot{\gamma}_{13}|$. There is a class of restricted flows called *viscometric flows*, which are motions equivalent to steady simple shearing. Tanner (2000) has show various viscometric kinematic fields where each fluid element is undergoing a steady simple shearing motion, with streamlines that are straight, circular, or helical. Each flow can be viewed as a relative sliding motion of a shear of inextensible material surfaces, which are called *slip surfaces.*

Another important class of fluid flow is the extensional flow (or, elongational flow), which refers to flow where the rate of deformation tensor is diagonal. For an incompressible fluid, one can define the following form (Dealy 1994):

$$\mathbf{D} = \dot{\varepsilon}_{11}\begin{bmatrix} 1 & 0 & 0 \\ 0 & a & 0 \\ 0 & 0 & -(1+a) \end{bmatrix}, \tag{2.16}$$

where $-0.5 < a < 1$, with $a = -0.5$ for uniaxial extension, $a = 0$ for planar extension, and $a = 1$ for equibiaxial extension.

For the uniaxial extensional flow, the generalized strain rate

$$\dot{\gamma} = \sqrt{3}\dot{\varepsilon}_{11}. \tag{2.17}$$

For the uniaxial extensional flow the extensional viscosity is defined as (Dealy 1994)

$$\eta_E = \frac{N_1}{\dot{\varepsilon}_{11}}, \tag{2.18}$$

where N_1 is the first normal stress difference (σ_{11}–σ_{22}). For a Newtonian fluid, its extensional viscosity is three times its shear viscosity, i.e.,

$$\eta_E = 3\eta. \tag{2.19}$$

Flows in injection molding are dominated by viscometric flows, but extensional flows are also encountered, for example, in the mid-surface of a center-gated cavity, or in a cavity having a sudden change in thickness, or in the "fountain flow". Calculations of flow through runners and gates usually need to consider the effect of extensional viscosity (Brincat et al. 1998; Gupta 2000).

2.1.3 Conservation Equations

We now consider the equations governing the fluid flow, namely the equations of conservation of mass, momentum and energy. Derivations of these equations in general form can be found in many text books. These derivations will not be repeated here.

2.1.3.1 Conservation of Mass

The differential equation expressing the conservation of mass, also known as the continuity equation, is

$$\frac{\partial \rho}{\partial t} + \frac{\partial}{\partial x_i}(\rho u_i) = 0, \tag{2.20}$$

or, equivalently,

$$\frac{\mathrm{D}\rho}{\mathrm{D}t} + \rho\frac{\partial u_i}{\partial x_i} = 0, \tag{2.21}$$

where ρ is the density field at time t. The density of a polymer melt depends on temperature and pressure. It is sometimes assumed that the fluid has a constant density then the equation of mass conservation reduces to

$$\frac{\partial u_i}{\partial x_i} = 0. \tag{2.22}$$

This is also known as the condition of incompressibility. Hence we also have $D_{ii} = 0 (tr\,\mathbf{D} = 0)$ for incompressible fluid.

However, the change of density cannot be neglected under the high level of pressure involved in processes such as injection molding.

2.1.3.2 Conservation of Momentum

The equations of the conservation of momentum, also called the equations of motion, are

$$\frac{\partial \sigma_{ij}}{\partial x_j} + \rho f_i = \frac{\partial}{\partial t}(\rho u_i) + \frac{\partial}{\partial x_j}(\rho u_i u_j), \tag{2.23}$$

where σ_{ij} is the total stress tensor and f_i is the body force per unit mass. If we account for the mass conservation equation, the equation can be written as

$$\frac{\partial \sigma_{ij}}{\partial x_j} + \rho f_i = \rho \left(\frac{\partial u_i}{\partial t} + u_j \frac{\partial u_i}{\partial x_j} \right). \tag{2.24}$$

The law of angular momentum balance requires that the stress tensor is symmetric

$$\sigma_{ij} = \sigma_{ji}. \tag{2.25}$$

This symmetry condition for the stress tensor will be used throughout this book.

For steady-state flows (here we consider the "steady-state" in the Eulerian viewpoint. In a steady Eulerian velocity field (which is not necessarily Lagrangian steady), $\partial u_i/\partial t = 0$. For the analysis of polymer melt flow occurring at a very low Reynolds number, the inertia term $\rho u_j \, \partial u_i/\partial x_j$ will usually be neglected. The body force effect is usually negligible too. This reduces the equations of motion to

$$\frac{\partial \sigma_{ij}}{\partial x_j} = 0. \tag{2.26}$$

In solving particular problems, suitable boundary conditions must be applied in particular circumstances. For the equations of motion, we distinguish between velocity boundary conditions and traction boundary conditions.

The velocity boundary condition is prescribed on the parts of boundary Γ_u:

$$u_i = U_i \text{ on } \Gamma_u. \tag{2.27}$$

This type of boundary condition is also called an *essential boundary condition* in numerical simulations.

The traction boundary condition is prescribed on the parts of boundary Γ_t:

$$\sigma_{ij} n_j = t_i \text{ on } \Gamma_t, \tag{2.28}$$

where t_i is the traction (or the total surface stress), and n_i is the outward unit vector normal to the boundary. The traction boundary condition is also called a *natural boundary condition* in numerical simulations.

2.1.3.3 Conservation of Energy

The energy conservation equation is often expressed in terms of temperature as

$$\rho c_p \frac{\mathrm{D}T}{\mathrm{D}t} + \frac{T}{\rho}\left(\frac{\partial \rho}{\partial T}\right)_P \frac{\mathrm{D}P}{\mathrm{D}t} = \tau_{ij} D_{ij} - \frac{\partial q_i}{\partial x_i} + \dot{Q}, \tag{2.29}$$

where T is the temperature, c_p is the specific heat at constant pressure, P is the pressure, $\tau_{ij}\,D_{ij}$ is the heat due to viscous dissipation, q_i is the heat flux vector and $\dot{Q}$ is the volumetric heat source.

For thermally isotropic materials, the heat flux vector is given by Fourier's law

$$q_i = -k\frac{\partial T}{\partial x_i}, \tag{2.30}$$

where k is the thermal conductivity. For thermally anisotropic materials, one expresses the heat conduction by the following non-Fourier form (Huilgol et al. 1992):

$$q_i = -k_{ij}\frac{\partial T}{\partial x_j}, \tag{2.31}$$

where k_{ij} is the thermal conductivity tensor.

For semicrystalline polymer flows, the release of latent heat due to crystallization should be taken into account as the heat source term in the energy equation, i.e.,

$$\dot{Q} = \rho \Delta H_c \frac{\mathrm{D}\chi}{\mathrm{D}t}, \tag{2.32}$$

where ΔH_c is the latent heat of fusion per unit mass for a perfect crystals, and χ is the degree of crystallinity.

In injection molding problems, the fixed boundary is the interface between the polymer and the metal mold walls. The temperature boundary condition can either be a temperature

$$T = T_w, \tag{2.33}$$

where T_w is specified, or a heat flow

$$k\frac{\partial T}{\partial n} = q_w, \tag{2.34}$$

where q_w is specified, and $\partial T/\partial n = (\partial T/\partial x_i)n_i$ is the temperature gradient in the direction normal to the boundary.

2.2 Constitutive Equations

2.2.1 Newtonian Fluids

A general form of the constitutive equation for the Newtonian fluid is given by

$$\sigma_{ij} = -P\delta_{ij} + \left[\mu_d - \frac{2}{3}\eta\right] D_{kk}\delta_{ij} + 2\eta D_{ij}, \tag{2.35}$$

where the scalar P is the *pressure*, η is the constant shear viscosity and μ_d is called the *dilatational viscosity*.

For compressible fluids the pressure can be identified with the *thermodynamic pressure* that depends on density and temperature; for incompressible fluids the pressure is simply regarded as another dynamical variable to be determined as part of the solution to a flow problem.

The dilatational viscosity is zero for ideal gases that are highly compressible. For incompressible fluids, $D_{kk} = 0$ and the term containing μ_d vanishes. Since the dilatational viscosity has no effect in the two extreme cases, we assume it is negligible for polymer flows. With $\mu_d = 0$, one can prove that pure volumetric change does not affect the total stress. To prove this, in Eq. 2.35 we replace the subscript indices *ij* by repeated indices *ii*, and note that $\delta_{ii} = 3$ and $D_{kk} = D_{ii}$ according to the summation convention. We obtain

$$\sigma_{ii} = -3P - 3 \times \frac{2}{3}\eta D_{ii} + 2\eta D_{ii} = -3P, \tag{2.36}$$

which shows that the trace of the total stress tensor is independent of the trace of the deformation rate tensor. Furthermore, the term of $(-2/3)\eta D_{kk}\delta_{ij}$, which does not dissipate energy, can be lumped into the pressure term. This lead to the familiar constitutive equation for Newtonian fluids:

$$\sigma_{ij} = -P\delta_{ij} + 2\eta D_{ij}. \tag{2.37}$$

It is customary to decompose the total stress tensor into two parts:

$$\sigma_{ij} = -P\delta_{ij} + \tau_{ij}, \tag{2.38}$$

where $\boldsymbol{\tau}_{ij}$ is called the *extra stress* tensor. As a result of being free of external moment, the extra stress tensor is symmetric:

$$\tau_{ij} = \tau_{ji}. \tag{2.39}$$

The extra stress is the stress arising from deformation of the material, and is related to the deformation by constitutive equations. For example, for the Newtonian fluid,

$$\tau_{ij} = 2\eta D_{ij}. \tag{2.40}$$

2.2.2 Generalized Newtonian Fluids

The choice of constitutive equations depends on the particular problems investigated. If the flow phenomena are dominated by the shear-rate dependent viscosity, it makes sense to use inelastic, or *generalized Newtonian* fluids, for which the extra stress tensor is proportional to the rate of deformation in the form

$$\tau_{ij} = 2\eta(\dot{\gamma})D_{ij}, \tag{2.41}$$

where $\eta(\dot{\gamma})$ is the shear viscosity as a function of the generalized strain rate $\dot{\gamma} = (2D_{ij}D_{ij})^{1/2}$. The shear viscosity can also depend on temperature and pressure. One can consider the effects based on the free-volume concept and using the time–temperature superposition approach in the calculation. This will be discussed later.

The difference between various models of the generalized Newtonian fluid is the expression used for the viscosity function $\eta(\dot{\gamma})$.

2.2.2.1 Power-law Model

This model has the form

$$\eta = m\dot{\gamma}^{n-1}, \tag{2.42}$$

where m and n are the *consistency constant* and the *power-law index*, respectively. Owing to its simplicity and the small number of material parameters, it was widely used in early injection modeling, see for example, Hieber and Shen (1980). However, the model does not correctly predict the zero-shear-rate viscosity at the low shear rate region—in this region the viscosity is unbounded. This serious limitation makes it become less attractive today for solving practical injection molding problems.

2.2.2.2 Carreau Model

The Carreau model (see Bird et al. 1987a) is expressed as

$$\frac{\eta - \eta_\infty}{\eta_0 - \eta_\infty} = \left[1 + (\lambda\dot{\gamma})^2\right]^{\frac{n-1}{2}}, \tag{2.43}$$

where η_0 is the zero-shear-rate viscosity, η_∞ is the upper limiting Newtonian viscosity, λ is a time constant, and n is the power-law index. The model has been successful in correlating experimental viscosity data. When $\eta_\infty = 0$ and at high shear rate, the model reduces to the power-law model with $m = \eta_0 \lambda^{n-1}$.

A well-known relationship between the zero-shear-rate viscosity and molecular weight M_w is that (Fox and Flory 1954)

$$\eta_0 \propto M_W^{3.4}. \tag{2.44}$$

2.2.2.3 Cross Model

The Cross model (Cross 1965) is given as

$$\eta = \frac{\eta_0}{1 + (\eta_0 \dot{\gamma}/\tau^*)^{1-n}}, \tag{2.45}$$

which combines a Newtonian region and a power -law shear thinning region. When $\dot{\gamma} \to 0$, it predicts the zero-shear-rate viscosity, η_0, at high shear rate, it predicts power-law behavior. In the equation, τ^* is a constant related to the shear stress at the transition between Newtonian and power-law behavior, and n is the power-law index, a measure of the degree of the shear-thinning behavior.

In comparing the Carreau and Cross models on a variety of commercial-grade polymer melts, Hieber and Chiang (1992) found that the Cross model provides a better overall fit for the shear-rate dependence. The Cross model has been widely used in injection molding modeling.

The generalized Newtonian fluid model has no memory, and no elasticity either; it predicts a zero first normal stress difference ($N_1 = \sigma_{11} - \sigma_{22} = \tau_{11} - \tau_{22} = 0$) in a simple shear flow. None of the above models are suitable for extensional flows, since they predict the extensional viscosity

$$\eta_E = 3\eta\left(\sqrt{3}\dot{\varepsilon}_{11}\right), \tag{2.46}$$

which is not accurate for polymer flows. Debbaut and Crochet (1988) proposed modifications to the generalized Newtonian fluid model for complex flows, where is a mixture of shear and extensional flow components. The basic idea is to let the viscosity depend on both the second and the third invariants of the rate of deformation tensor, i.e., $\tau_{ij} = 2\eta(II_D, III_D)D_{ij}$. However, since the third invariant is always zero in plane flows, this approach is of limited utility.

2.2.3 Linear Viscoelastic Models

The linear viscoelastic model assumes that the stress at the current time depends not only on the current strain, but on the past strains as well. It also assumes a linear superposition. Its general form reads

$$\tau_{ij}(t) = 2 \int_{-\infty}^{t} G(t - t')D_{ij}(t')\mathrm{d}t', \tag{2.47}$$

where $G(t)$ is the relaxation modulus.

The inelastic liquid is recovered with the relaxation modulus function being set to

$$G(t) = \eta\delta(t), \tag{2.48}$$

where $\delta(t)$ is the Dirac delta function, or the impulse function. A key property of the function is that for any function $f(x)$,

$$\int_{-\infty}^{\infty} f(x)\delta(a - x)\mathrm{d}x = f(a), \tag{2.49}$$

hence

$$\tau_{ij}(t) = 2 \int_{-\infty}^{t} \eta\delta(t - t')D_{ij}(t')\mathrm{d}t' = 2\eta D_{ij}(t). \tag{2.50}$$

The most often used relaxation modulus function is the multi-mode Maxwell memory function:

$$G(t) = \sum_{i=1}^{N} G_i \exp(-t/\lambda_i), \tag{2.51}$$

which consists of a discrete relaxation spectrum $\{G_i, \lambda_i\}$. This function describes a fading memory. Thus, the linear viscoelastic model shows that a strain at a distant past contributes less to the stress than a more recent strain.

2.2.4 Viscoelastic Fluid Models

The generalized Newtonian fluid and the linear viscoelastic models are not capable of describing most of the essential non-Newtonian properties of polymers in complex flows. For the injection molding application, although the generalized Newtonian fluid model could provide accurate results in pressure development and flow front advancement during cavity filling, it may predict incorrect results when applied to regions having a sudden change in geometry, and, even in the regions dominated by the viscometric flow, the inelastic analysis cannot provide information about frozen-in stresses and flow-induced micro-structures. Thus, we shall review some viscoelastic models here. More complete details of viscoelastic fluid models can be found in several books (e.g., Bird et al. 1987a, b; Tanner 2000; Huilgol and Phan-Thien 1997).

2.2.4.1 Linear Elastic Dumbbell Model

The simplest model for dilute polymer solutions is to idealize the polymer molecule as an elastic dumbbell consisting of two beads connected by a Hookean spring immersed in a viscous fluid (Fig. 2.1). The spring has an elastic constant H_0. Each bead is associated with a frictional factor ζ and a negligible mass. If the instantaneous locations of the two beads in space are $\mathbf{r}_1$and $\mathbf{r}_2$, respectively, then the end-to-end vector, $\mathbf{R} = \mathbf{r}_2 - \mathbf{r}_1$, describes the overall orientation and the internal conformation of the polymer molecule. The polymer-contributed stress tensor can be related to the second-order moment of $\mathbf{R}$. There are two expressions namely the Kramers expression and the Giesekus expression, respectively (Bird et al. 1987b):

$$\text{Kramers: } \tau_{ij} = -n_0 k_B T \delta_{ij} + n_0 \langle F_i^{(c)} R_j \rangle, \tag{2.52}$$

$$\text{Giesekus: } \tau_{ij} = -\frac{n_0 \zeta}{4} \frac{\Delta \langle R_i R_j \rangle}{\Delta t}, \tag{2.53}$$

where n_0 is the number of molecules per unit volume, $k_B = 1.38 \times 10^{-23}$ J/K is the Boltzmann constant, T is the temperature, $F_i^{(c)}$ is the tension in the connector, and the angular bracket $\langle\,\rangle$ denotes the ensemble average with respect to the probability density function of the variable concerned. The derivative $\Delta/\Delta t$ is called the upper convected derivative defined as

$$\Delta(\,)_{ij}\big/\Delta t = \mathrm{D}(\,)_{ij}\big/Dt - L_{ik}(\,)_{kj} - L_{jk}(\,)_{ki},$$

where L_{ij} is the velocity gradient. The upper convected derivative was introduced by Oldroyd (1950) to guarantee the objectivity of constitutive equations.

For the Hookean spring, the connector force $F_i^{(c)} = H_0 R_i$ and hence the Kramers expression becomes

$$\tau_{ij} = -n_0 k_B T \delta_{ij} + n_0 H_0 \langle R_i R_j \rangle. \tag{2.54}$$

One can simply eliminate $\boldsymbol{\tau}_{ij}$ from Eqs. 2.52 and 2.54, and obtain the following equation for $\langle R_i R_j \rangle$:

$$\lambda \frac{\Delta \langle R_i R_j \rangle}{\Delta t} + \langle R_i R_j \rangle = \frac{k_B T}{H_0} \delta_{ij}. \tag{2.55}$$

where $\lambda = \zeta/4H_0$ is called the Rouse relaxation time. Using the Kramers formulation, we can define a stress tensor S_{ij} as

$$S_{ij} \equiv \tau_{ij} + n_0 k_B T \delta_{ij}, \tag{2.56}$$

and hence, from Eq. 2.54

$$S_{ij} = n_0 H_0 \langle R_i R_j \rangle. \tag{2.57}$$

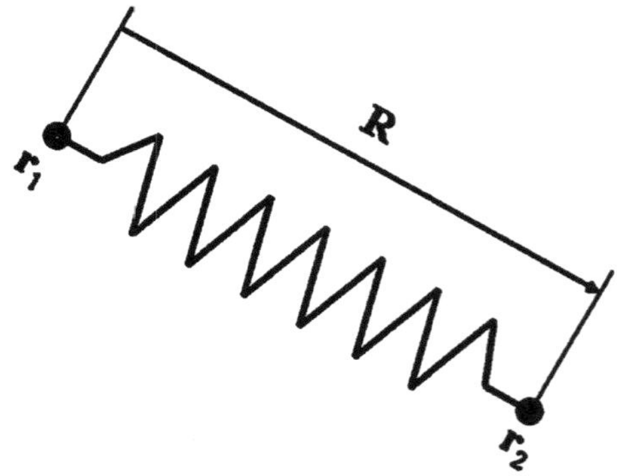

Fig. 2.1 Elastic dumbbell model

By eliminating $\langle R_i\, R_j \rangle$ from Eqs. 2.57 and 2.55, one has

$$\lambda \frac{\Delta S_{ij}}{\Delta t} + S_{ij} = G\delta_{ij}, \tag{2.58}$$

where $G = n_0 k_B T$. Then, since

$$\frac{\Delta \delta_{ij}}{\Delta t} = \frac{\mathrm{D}\delta_{ij}}{\mathrm{D}t} - L_{ik}\delta_{kj} - L_{jk}\delta_{ki} = -(L_{ij} + L_{ji}) = -2D_{ij},$$

by substituting Eq. 2.56 into 2.58, we obtain

$$\lambda \frac{\Delta \tau_{ij}}{\Delta t} + \tau_{ij} = 2\eta D_{ij}, \tag{2.59}$$

where $\eta = \lambda G$ is the polymer viscosity.

This equation is also called the Upper-Convected Maxwell (UCM) model. In a steady state simple shear flow, the model predicts a first normal stress difference which is proportional to the square of shear rate, and a zero second normal stress. It however does not predict shear thinning observed in most polymeric fluids. A more serious limitation is that in a uniaxial extensional flow, the model exhibits a singularity in extensional viscosity at $\lambda\dot{\varepsilon} = 0.5$. This is an unrealistic feature of the model. In the area of numerical modeling of viscoelastic flows, researchers were long frustrated by a difficulty called the "high Weissenberg number problem"—there was an upper limit on the Weissenberg number, above which the numerical simulation failed (Tanner 1982; Crochet et al. 1984; Keunings 1986). The problem was first confronted when using the UCM model to solve complex flow problems. Therefore, the usefulness of the UCM model for practical problems is limited. The model is mainly useful for illustrative purposes.

2.2.4.2 Finite Extensible Non-linear Elastic Dumbbell Model

The unrealistic behavior of the UCM results from the fact that the Hookean spring allows extensions to go to infinity. A way of improvement is to use a more realistic force law in the model. Warner (1972) replaced the Hookean spring constant H_0 by

$$H = \frac{H_0}{1 - (R/L)^2}, \tag{2.60}$$

where $R = (R_i \ R_i)^{1/2}$ is the extension and L is the maximum extension of the dumbbell. This spring will get stiffer as it is extended, and it cannot be extended beyond a length L.

For finite values of L, however, it is impossible to obtain a closed-form constitutive equation unless an approximation is made. A well-known one was made by Peterlin (see, for example, Keunings 1997)

$$\left\langle \frac{R_i R_j}{1 - R^2/L^2} \right\rangle \approx \frac{\langle R_i R_j \rangle}{1 - \langle R^2 \rangle / L^2}. \tag{2.61}$$

The dumbbell is thus called the FENE-P dumbbell ("FENE" stands for "Finite Extendable Nonlinear Elastic", and "P" stands for Peterlin, who first suggested this simplification). With this approximation, the Kramers formulation becomes

$$\tau_{ij} = -n_0 k_B T \delta_{ij} + \frac{n_0 H_0 \langle R_i R_j \rangle}{1 - \langle R^2 \rangle / L^2} \tag{2.62}$$

If we define a microstructural tensor by

$$c_{ij} = \frac{H_0 \langle R_i R_j \rangle}{k_B T}, \tag{2.63}$$

and let

$$b = \frac{L^2 H_0}{k_B T}, \tag{2.64}$$

Equation 2.62 can be rewritten as

$$\tau_{ij} = G\left[\left(1 - \frac{c_{kk}}{b}\right)^{-1} c_{ij} - \delta_{ij} \right], \tag{2.65}$$

where $G = n_0 k_B T$. We also have the Giesekus formulation, valid for all dumbbells

$$\tau_{ij} = -n_0 \lambda H_0 \frac{\Delta \langle R_i R_j \rangle}{\Delta t}, \tag{2.66}$$

which yields

$$\tau_{ij} = -G\lambda \frac{\Delta c_{ij}}{\Delta t}. \tag{2.67}$$

Thus $\boldsymbol{\tau}_{ij}$ may be eliminated from Eqs. 2.65 and 2.67, leading to

$$\lambda \frac{\Delta c_{ij}}{\Delta t} + \left(1 - \frac{c_{kk}}{b}\right)^{-1} c_{ij} = \delta_{ij}. \tag{2.68}$$

Equations 2.65 and 2.68 are the required equations. One can also eliminate c_{ij} from the above equations to obtain a macroscopic constitutive equation in term of the polymer stress tensor, see Tanner (1975).

The FENE-P model is able to predict qualitatively many effects in dilute polymer solutions such as shear thinning, first normal stress difference, and a "stiffer" response in extensional viscosity. The model is an improvement on the Maxwell model, but it does not always behave well in transient flows.

2.2.4.3 Rigid Dumbbell Model

There are cases where a rigid-rod microstructure needs to be considered. The simplest model for this is the rigid dumbbell model (Fig. 2.2), where the two beads, located at $\mathbf{r}_1$ and $\mathbf{r}_2$, keep a constant distance R apart. The end-to-end vector is $\mathbf{R} = \mathbf{r}_2 - \mathbf{r}_1 = R\mathbf{p}$, where $\mathbf{p}$ is a unit vector directed from bead 1 to bead 2. The relative velocity of two unrestrained beads is $\dot{\mathbf{R}} = \mathbf{L} \cdot \mathbf{R} \equiv R\mathbf{L} \cdot \mathbf{p}$, where $\mathbf{L}$ (or using the notation L_{ij}) is the velocity gradient tensor of the surrounding liquid. The velocity component along the $\mathbf{p}$-vector is $\dot{\mathbf{R}} \cdot \mathbf{p} = R\mathbf{L}:\mathbf{pp}$, thus the relative velocity vector in the $\mathbf{p}$-vector direction is $R\mathbf{L}:\mathbf{ppp}$. Since the two beads are constrained by a rigid link that cannot be stretched, the stretching component $R\mathbf{L}:\mathbf{ppp}$ must be subtracted from $R\mathbf{L} \cdot \mathbf{p}$. We thus have the relative velocity of the two restrained beads given by

$$\dot{\mathbf{R}} = R\dot{\mathbf{p}} = R\mathbf{L} \cdot \mathbf{p} - R\mathbf{L}:\mathbf{ppp}. \tag{2.69}$$

Then we obtain the equation for the motion of the unit vector $\mathbf{p}$:

$$\dot{\mathbf{p}} = \mathbf{L} \cdot \mathbf{p} - \mathbf{L}:\mathbf{ppp} \text{ or } \dot{p}_i = L_{ij}p_j - L_{kl}p_k p_l p_i. \tag{2.70}$$

Brownian motion can also be considered. Let the Brownian force be $\mathbf{F}^{(b)}$, represented by some white noise, which has zero mean and delta correlation function:

$$\langle \mathbf{F}^{(b)}(t+s)\mathbf{F}^{(b)}(t) \rangle = 2D_r\delta(s)\mathbf{I}, \tag{2.71}$$

where D_r is the strength of the white noise, called the rotary diffusion coefficient, $\delta(s)$ is the impulse or delta Dirac function.

The part of $\mathbf{F}^{(b)}$ parallel to $\mathbf{p}$ is $(\mathbf{F}^{(b)} \cdot \mathbf{p})\mathbf{p} = \mathbf{pp} \cdot \mathbf{F}^{(b)}$. We need to consider only the effect of the part of the force normal to $\mathbf{p}$. Therefore we subtract the parallel part from $\mathbf{F}^{(b)}$ and obtain the reduced Brownian force of $(\mathbf{I} - \mathbf{pp}) \cdot \mathbf{F}^{(b)}$ on the space orthogonal to $\mathbf{p}$. Adding the result to Eq. 2.70 yields

$$\dot{p}_i = L_{ij}p_j - L_{kl}p_k p_l p_i + (\delta_{ij} - p_i p_j)F_j^{(b)}. \tag{2.72}$$

We want to derive an evolution equation in terms of the ensemble average $\langle \mathbf{pp} \rangle$. For this purpose, let

$$Q_{ij} = p_i p_j. \tag{2.73}$$

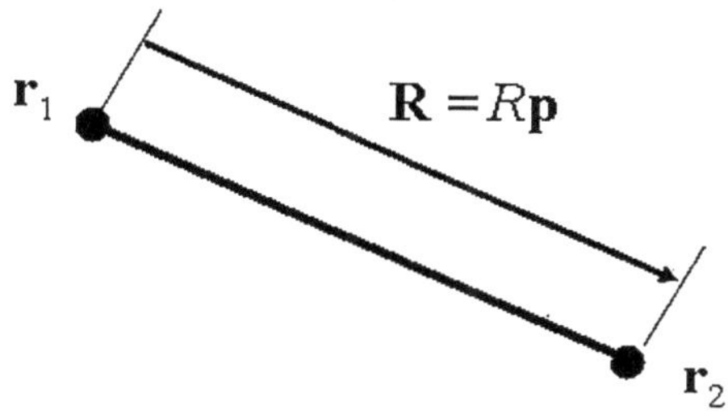

Fig. 2.2 Rigid dumbbell model

So

$$\begin{aligned}\frac{\mathrm{D}Q_{ij}}{\mathrm{D}t} &= \frac{\partial Q_{ij}}{\partial p_k}\dot{p}_k \\ &= \frac{\partial Q_{ij}}{\partial p_k}\left[L_{km}p_m - L_{mn}p_m p_n p_k + (\delta_{km} - p_k p_m)F_m^{(b)}\right].\end{aligned} \tag{2.74}$$

Thus,

$$\begin{aligned}\frac{\mathrm{D}\langle Q_{ij}\rangle}{\mathrm{D}t} =& L_{km}\left\langle p_m \frac{\partial Q_{ij}}{\partial p_k}\right\rangle - L_{mn}\left\langle p_m p_n p_k \frac{\partial Q_{ij}}{\partial p_k}\right\rangle \\ &+ \left\langle \frac{\partial Q_{ij}}{\partial p_k}F_k^{(b)}\right\rangle - \left\langle \frac{\partial Q_{ij}}{\partial p_k}p_k p_m F_m^{(b)}\right\rangle.\end{aligned} \tag{2.75}$$

Using Eq. 2.71, the last two terms in Eq. 2.75 become (Phan-Thien and Zheng 1997)

$$\left\langle \frac{\partial Q_{ij}}{\partial p_k}F_k^{(b)}\right\rangle - \left\langle \frac{\partial Q_{ij}}{\partial p_k}p_k p_m F_m^{(b)}\right\rangle = D_r\left\langle (\delta_{km} - p_k p_m)\frac{\partial^2 Q_{ij}}{\partial p_k \partial p_m} - 2p_k \frac{\partial^2 Q_{ij}}{\partial p_k}\right\rangle, \tag{2.76}$$

so that

$$\begin{aligned}\frac{\mathrm{D}\langle Q_{ij}\rangle}{\mathrm{D}t} =& L_{km}\left\langle p_m \frac{\partial Q_{ij}}{\partial p_k}\right\rangle - L_{mn}\left\langle p_m p_n p_k \frac{\partial Q_{ij}}{\partial p_k}\right\rangle \\ &+ D_r\left\langle (\delta_{km} - p_k p_m)\frac{\partial^2 Q_{ij}}{\partial p_k \partial p_m} - 2p_k \frac{\partial Q_{ij}}{\partial p_k}\right\rangle.\end{aligned} \tag{2.77}$$

Note that

$$\frac{\partial Q_{ij}}{\partial p_k} = \delta_{ik}p_j + \delta_{jk}p_i - 2p_i p_j p_k \tag{2.78}$$

By substituting Eqs. 2.78 and 2.73 into Eq. 2.77, we have

$$\frac{D\langle p_i p_j \rangle}{Dt} = L_{im}\langle p_m p_j \rangle + L_{jm}\langle p_m p_i \rangle \\ - 2L_{mn}\langle p_m p_n p_i p_j \rangle + 2D_r\left(\delta_{ij} - 3\langle p_i p_j \rangle\right), \quad (2.79)$$

which can be rewritten as

$$\Lambda\left(\frac{\Delta\langle p_i p_j \rangle}{\Delta t} + 2L_{mn}\langle p_m p_n p_i p_j \rangle\right) + \langle p_i p_j \rangle = \frac{1}{3}\delta_{ij}, \quad (2.80)$$

where $\Lambda \equiv 1/(6D_r)$ and $\Delta/\Delta t$ is the upper convective derivative. The Kramers expression of the stress can be found to be (Bird et al. 1987b)

$$\tau_{ij} = -n_0 k_B T \delta_{ij} + 3n_0 k_B T\left(\langle p_i p_j \rangle + 2\Lambda L_{mn}\langle p_m p_n p_i p_j \rangle\right). \quad (2.81)$$

2.2.4.4 Phan-Thien-Tanner (PTT) Model and Giesekus Model

The elastic and the FENE dumbbell models lack interactions between different parts of the chain, and between different chains (such as cross-linking or entanglements in a concentrated polymer system). There are two more sophisticated theories: network and reptation theories. The Phan-Thien–Tanner (PTT) model (Phan-Thien and Tanner 1977) was derived based on network theory (Lodge 1964). This approach is concerned with a balance between the creation and destruction of strands in the network of long chain polymer molecules. The result for the one-relaxation-tome form is

$$\lambda\frac{\delta S_{ij}}{\delta t} + H_{ik}S_{kj} = G_1\delta_{ij}. \quad (2.82)$$

where S_{ij} is an extra stress tensor, H_{ik} is a network destruction function, and G_1 is a network creation function, λ is the relaxation time. The symbol $\delta/\delta t$ represents the Gordon–Schowalter convected derivative defined as (Gordon and Schowalter 1972)

$$\frac{\delta S_{ij}}{\delta t} \equiv \frac{DS_{ij}}{Dt} - \mathscr{L}_{ik}S_{kj} - \mathscr{L}_{jk}S_{ki} \quad (2.83)$$

where $\mathscr{L}_{ij} = L_{ij} - \xi D_{ij}$ is an effective velocity gradient tensor, and ξ is call the slip parameter accounting for the non-affine motion of the network strands. If we define $H_{ik} = H_1\delta_{ik}$, $G = \eta/\lambda$ and let

$$S_{ij} = \frac{G}{1-\xi}\delta_{ij} + \tau_{ij}, \quad (2.84)$$

and

$$G_1 = \frac{G}{1-\xi}H_1, \quad (2.85)$$

then, substitution of 2.84 and (2.85) to (2.82) gives

$$\lambda \frac{\delta \tau_{ij}}{\delta t} + H_1 \tau_{ij} = 2\eta D_{ij}. \tag{2.86}$$

Phan-Thien and Tanner (1977) have suggested two empirical forms for H_1:

$$H_1 = \begin{cases} 1 + \frac{\lambda \varepsilon}{\eta} \tau_{kk} \\ \exp\left(\frac{\lambda \varepsilon}{\eta} \tau_{kk}\right) \end{cases},$$

where ε is a parameter governing the extensional flow response.

The PTT model predicts shear thinning and first normal stress difference for both steady and transient shearing. It also predicts a non-zero (negative) second normal stress difference in simple shear flow: $N_2 = -\xi N_1/2$. The main advantage of the PPT model is that it predicts reasonable extensional flow behavior at all extensional rates.

When $\varepsilon = 0$ and $\xi = 0$, the PTT model reduces to the UCM model.

If $\xi = 0$, and $H_{ik} = a_G\, S_{ik}/G + (1 - a_G \delta_{ik})$, then we find:

$$\lambda \frac{\Delta \tau_{ij}}{\Delta t} + \left(\frac{a_G}{G} \tau_{ik} + \delta_{ik}\right) \tau_{kj} = 2\eta D_{ij}, \tag{2.87}$$

which is the Giesekus model (Giesekus 1982, 1983), with α_G being a constant called the Giesekus constant, ranging from 0 to 1. When $\alpha_G = 0$ one recovers the UCM model.

2.2.4.5 The eXtended Pom–Pom (XPP) Model

Because of entanglements, a long molecule is not allowed to cross any of the surrounding molecules, but it can move in between in a slow snake-like manner. This view of molecular motion has lead to the *reptation model* of de Gennes (1979) and Doi and Edwards (1986). In this model, a polymer chain is assumed to be constrained in a "tube" created by the surrounding chains and deformed with the tube cooperatively. The so-called "Pom-Pom" model, originally proposed by McLeish and Larson (1998), is based on the reptation theory. The model considers a class of branched molecules as a molecular chain consisting of a backbone, to which a number of arms (q) are attached at the ends. The motion of the arms and backbones in "tubes" are then considered. The model consists of two equations: one for the orientation and one for the stretch. The relaxation times for the orientation and the stretch are separated. This original Pom-Pom model has some weakness, such as discontinuous solutions for steady state elongational flows, unbounded orientation equation for high strain rates, and zero second normal stress difference in shear. To overcome these limitations, Verbeeten et al. (2001, 2002) proposed an improved Pom-Pom model and called it the eXtended Pom-Pom model (XPP), which can be written as

$$\lambda \frac{\Delta \tau_{ij}}{\Delta t} + K_{ik}\tau_{kj} = 2\eta D_{ij}, \tag{2.88}$$

with

$$K_{ik} = \frac{a_G}{G}\tau_{ik} + \Gamma\delta_{ik} + G(\Gamma - 1)\tau_{ik}^{-1}, \tag{2.89}$$

where a_G is the Giesekus constant, and the auxiliary scalar valued function Γ is defined by

$$\Gamma = \frac{2\lambda}{\lambda_s}\exp\left(\frac{2(\Lambda - 1)}{q}\right)\left(1 - \frac{1}{\Lambda}\right) + \frac{1}{\Lambda^2}\left(1 - \frac{\alpha_G \tau_{ij}\tau_{ij}}{3G^2}\right), \tag{2.90}$$

which involves two relaxation times: λ is the usual relaxation time characterizing the relaxation of the backbone orientation, and λ_s is the relaxation time of the tube stretch. The parameter Λ is the backbone tube stretch given by

$$\Lambda = \left(1 + \frac{\tau_{kk}}{3G}\right)^{1/2}. \tag{2.91}$$

Tanner and Nasseri (2003) and Tanner (2006) have shown that the reptation-based XPP model can also be derived from the network viewpoint and Eq. 2.88 becomes a special case of the general network model.

The XPP model has been used by Bogaerds et al. (2004) to study the viscoelastic instability in injection molding.

2.3 Time–Temperature Superposition

Viscoelastic functions depend on both temperature and time. For many polymers, the logarithmic plot of a viscoelastic function at the temperature T may be obtained from that at the temperature T_0 by shifting the curve along the logarithmic time axis by the amount of log a_T (T). This procedure is called time–temperature superposition. The ability to superpose viscoelastic data is known as thermorheological simplicity. Thermorheological simplicity demands that all the molecular mechanisms involved in the relaxation process have the same temperature dependencies.

The time–temperature shift factor a_T (T) has been successfully fitted by the WLF (Williams–Landel–Ferry) empirical equation (Williams et al. 1955; Ferry 1980):

$$\log a_T(T) = \frac{-C_1^g(T - T_g)}{C_2^g + T - T_g}, \tag{2.92}$$

Although empirical in nature, the WLF equation has a theoretical justification based on the free-volume idea, as described in Ferry (1980). At first, the WLF constants C_1^g and C_2^g were thought to be universal for all polymers. The universal values are $C_1^g = 17.44$, $C_2^g = 51.6$ K if T_g, is chosen as the reference temperature. For practical problems, the "universal values" should only be used in the absence of relevant experimental data.

If some experimental data for a_T are available, one may use an arbitrary reference temperature T_0 to replace T_g and rewrite the equation as

$$\log a_T(T) = \frac{-C_1^0(T - T_0)}{C_2^0 + T - T_0}, \tag{2.93}$$

where C_1^0 and C_2^0 are experimentally determined constants, The calculation can be done by rearranging the WLF equation in the following form:

$$-\frac{T - T_0}{\log a_T} = \frac{C_2^0}{C_1^0} + \frac{1}{C_1^0}(T - T_0). \tag{2.94}$$

If the material data follow the WLF equation, then a plot of $-(T - T_0)/\log a_T$ versus $(T - T_0)$ should be a straight line, and C_1^0 and C_2^0 can be worked out from the slope and intercept.

Alternatively, one may set the two constants C_1^0 and C_2^0 to 8.86 and 101.6 K, respectively, and then choose T_0 to give a best fit to the experimental data.

Pressure (P) also has an effect on the viscoelastic response. Again, if all the molecular mechanism involved in the relaxation process has the same pressure dependencies, one may modify the WLF equation to take into account the pressure effect (Moonan and Tschoegl 1985):

$$\log a_T(T, P) = \frac{-C_1^P[T - T_0 - \theta(P)]}{C_2^P(P) + T - T_0 - \theta(P)}, \tag{2.95}$$

where

$$\theta(P) = C_3^P(P)\ln\frac{1 + C_4^P P}{1 + C_4^P P_0} - C_5^P(P)\ln\frac{1 + C_6^P P}{1 + C_6^P P_0}, \tag{2.96}$$

in which P_0 is a reference pressure, Parameter $C_n^P (n = 1, 2, \ldots 6)$ can be determined from relaxation experiments at high pressure.

The WLF equation is widely applicable to amorphous polymers in the range $T_g < T < T_g + 100$ (°C or K). Kaelble (1985) suggested a modification to extend this range below T_g:

$$\log a_T(T) = \frac{-C_1^g(T - T_g)}{C_2^g + |T - T_g|}. \tag{2.97}$$

For higher temperatures ($T > T_g + 100$°C), or where T_g is irrelevant, better results are obtained with the Arrhenius equation

$$\ln a_T(T) = \frac{E_a}{R_g}\left(\frac{1}{T} - \frac{1}{T_0}\right), \tag{2.98}$$

where E_a is the activation energy and R_g is the gas constant.

Let us consider a material undergoing stress relaxation. The shift factor a_T (T) describes the change of material time-scale with temperature. When temperature rises, the material time-scale shortens so that the relaxation proceeds faster, as if the material has an "internal time clock"—one unit of material time is equivalent to a_T (T) units of observe time t. That is, the relaxation process of the material is calculated in a pseudo-time t/a_T (T). Consider a material undergoing stress relaxation. The relaxation speed is determined by an internal time-scale (or clock) within the material. As temperature rises, so does the amount of molecular motion occurring in one unit of an observation time, and the material's time scale shortens so that the relaxation proceeds faster. The shift factor a_T gives a shift of material time-scale with temperature.

Let G (t, T) be the stress relaxation modulus at temperature T, and let G (t, T_0) be the modulus at a reference temperature T_0. We have

$$G(t, T) = b_T G(t/a_T, T_0), \tag{2.99}$$

where the factor b_T is called the vertical shift factor, given by

$$b_T = \frac{\rho T}{\rho_0 T_0}, \tag{2.100}$$

where ρ is the density at T, and ρ_0 is the density at T_0. Since increasing the temperature reduces the density, the value of b_T is usually nearly unity and can often be ignored.

For the zero-shear-rate viscosity $\eta_0(T)$, we have

$$\begin{aligned} \eta_0(T) &= \int_0^\infty G(t, T)dt = \int_0^\infty G(t/a_T, T_0)dt \\ &= a_T \int_0^\infty G(\bar{t}, T_0)d\bar{t} = a_T \eta_0(T_0). \end{aligned} \tag{2.101}$$

Similarly, for the relaxation time $\lambda(T)$, we have

$$\lambda(T) = a_T \lambda(T_0). \tag{2.102}$$

The time–temperature superposition principle can be incorporated into isothermal constitutive equations (including the generalized Newtonian fluid model) to solve non-isothermal flow problems. If we define

$$\bar{t} = t/a_T, \quad \bar{\mathbf{L}} = a_T \mathbf{L} \text{ (and hence } \bar{\mathbf{D}} = a_T \mathbf{D}, \quad \bar{\dot{\gamma}} = a_T \dot{\gamma}), \tag{2.103}$$

then we have

$$\boldsymbol{\tau}(t, \mathbf{L}, T) = \boldsymbol{\tau}(t/a_T, a_T \mathbf{L}, T_0). \tag{2.104}$$

As an example, consider the incompressible Cross equation at temperature T_0:

$$\tau_{ij}(D_{ij}, T_0) = \frac{2\eta_0(T_0)D_{ij}}{1 + [\eta_0(T_0)\dot{\gamma}/\tau^*]^{1-n}}. \tag{2.105}$$

For the stress τ_{ij} at temperature T. the equation can be written as

$$\tau_{ij}(\bar{D}_{ij}, T) = \frac{2\eta_0(T_0)\bar{D}_{ij}}{1 + [\eta_0(T_0)\bar{\dot{\gamma}}/\tau^*]^{1-n}}, \tag{2.106}$$

which is exactly the same form of equation as it is at temperature T_0, with $\bar{D}_{ij} = a_T D_{ij}$, and $\bar{\dot{\gamma}} = a_T\dot{\gamma}$. This implies that the response of the fluid at temperature T is the same as the response of the fluid at temperature T_0, but at a reduced shear rate $a_T\dot{\gamma}$.

Similarly, for the UCM model at temperature T_0

$$\lambda_0 \frac{\Delta\tau_{ij}}{\Delta t} + \tau_{ij} = 2\eta D_{ij}, \tag{2.107}$$

one finds the following equation for the relation at temperature T:

$$\lambda_0 \frac{\Delta\tau_{ij}}{\Delta \bar{t}} + \tau_{ij} = 2\eta \bar{D}_{ij}, \tag{2.108}$$

with $\bar{t} = t/a_T$ and $\bar{D}_{ij} = a_T D_{ij}$.

Morland and Lee (1960) showed how to incorporate time–temperature shifting into linear viscoelastic boundary-value problems. For the purpose a pseudo-time $\xi(t)$ is introduced such that the amount of time that passes during an interval $d\xi$ is given by dt/a_T. Then we have

$$\xi(t) = \int_0^t \frac{dt'}{a_T(T(t'))}. \tag{2.109}$$

In the isothermal case the linear viscoelastic equation can be written as

$$\tau_{ij}(t) = 2\int_{-\infty}^{t} G(t - t')D_{ij}(t')dt'. \tag{2.110}$$

In the non-isothermal case it can be reformulated in terms of ξ:

$$\tau_{ij}(t) = 2\int_{-\infty}^{t} G(\xi(t) - \xi(t'))D_{ij}(\xi(t'))dt'. \tag{2.111}$$

This equation is useful for prediction of residual stresses of injection-molded polymer products.

2.4 The Pressure–Volume–Temperature (PVT) Relation

For thermoplastic polymers, the equation of state is of the type $v = v(P, T)$, and is usually represented as a PVT diagram that gives the specific volume, v, as a function of temperature T and pressure P. The most commonly used equation is the Tait equation (van Krevelen 1976; Zoller et al. 1989; Zoller and Fakhreddine 1994):

$$v(T,P) = v_0(T)\left[1 - C\ln\left(1 + \frac{P}{B(T)}\right)\right] + v_t(T,P), \qquad (2.112)$$

where $C = 0.0894$ is a constant, considered to be universal. $v_0(T)$ is given by

$$v_0(T) = \begin{cases} b_1^{(s)} + b_2^{(s)}(T - b_5), & \text{if} \quad T \leq T_{\text{trans}} \\ b_1^{(m)} + b_2^{(m)}(T - b_5), & \text{if} \quad T > T_{\text{trans}} \end{cases}, \qquad (2.113)$$

where the superscripts (m) and (s) represent molten state and solid state of the polymer. $B(T)$ is given by

$$B(T) = \begin{cases} b_3^{(s)} \exp[-b_4^{(s)}(T - b_5)], & \text{if} \quad T \leq T_{\text{trans}} \\ b_3^{(m)} \exp[-b_4^{(m)}(T - b_5)], & \text{if} \quad T > T_{\text{trans}} \end{cases}, \qquad (2.114)$$

and, for amorphous polymers, $v_t(P, T) = 0$, while for semi-crystalline polymers,

$$v_t(T,P) = \begin{cases} b_7 \exp[b_8(T - b_5) - b_9 P] & \text{if} \quad T \leq T_{\text{trans}} \\ 0 & \text{if} \quad T > T_{\text{trans}} \end{cases}, \qquad (2.115)$$

where T_{trans} is the transition temperature, assumed to be a linear function of pressure, i.e.,

$$T_{\text{trans}} = b_5 + b_6 P. \qquad (2.116)$$

Differentiating the equation allows one to obtain the isothermal compressibility, κ, and the coefficient of volume expansion, α_V:

$$\kappa = \frac{1}{v}\left(\frac{\partial v}{\partial P}\right) = -\frac{1}{\rho}\left(\frac{\partial \rho}{\partial P}\right), \qquad (2.117)$$

and

$$\alpha_V = -\frac{1}{v}\left(\frac{\partial v}{\partial T}\right) = \frac{1}{\rho}\left(\frac{\partial \rho}{\partial T}\right). \qquad (2.118)$$

Figure 2.3 shows a typical PVT diagram for semi-crystalline materials, characterized by three basic zones: melt zone, transition zone and solid zone. Amorphous materials display no sudden transition from melt to solid.

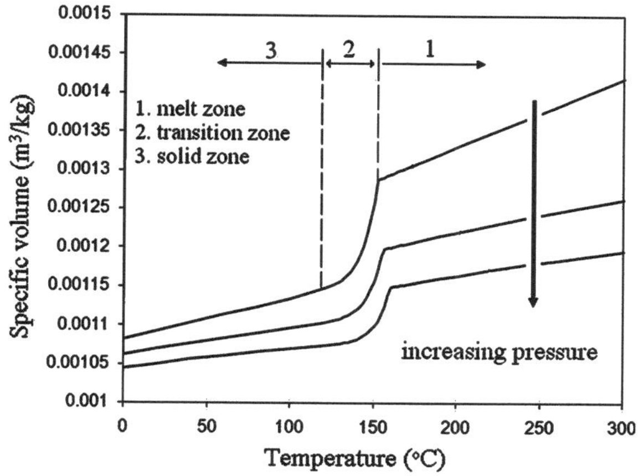

Fig. 2.3 PVT diagram for semi-crystalline material (polypropylene)

2.5 Lubrication Approximation

Creeping flows in slowly varying, relatively narrow gaps are often encountered in polymer processing, e.g., injection molding in particular. These kinds of flow are usually handled with the "lubrication approximation". Choose a coordinate system $(x_1, x_2, x_3,)$ such that the x_3-axis lies normal to the channel wall surfaces, and the x_1 and x_2 axes are tangential to these surfaces. The gap h in the fluid contact area will be very small compared with the characteristic length L in the x_1 and x_2 direction, and is a function of x_1 and x_2 only. By "slowly varying" we mean that $\partial h/\partial x_1 \ll 1$ and $\partial h/\partial x_2 \ll 1$. Under these conditions, the flow has the following characteristics:

1. The velocity gradient components $\partial u/\partial x_1$ and $\partial u/\partial x_2$ are negligible comparing to $\partial u/\partial x_3$.
2. At least for isothermal and pressure-independent Newtonian fluids, the pressure may be regarded as a function of x_1 and x_2 only, and thus $\partial P/\partial x_3 = 0$.
3. Assuming that stress components τ_{11} and τ_{13} are comparable in magnitude, then one has $\partial \tau_{11}/\partial x_1 \ll \partial \tau_{13}/\partial x_3$.
4. The Reynolds number is very small and the flow remains laminar.

A typical example of the lubrication flow is a Newtonian flow in a slider bearing where a moving wall drags the fluid with a velocity **U**. Taking the above-mentioned assumptions into account, the equation of motion governing the bearing flow becomes simply

$$\frac{\partial \tau_{13}}{\partial x_3} = \frac{\partial P}{\partial x_1}, \quad \tau_{13} = \eta \frac{\partial u_1}{\partial x_3}, \tag{2.119}$$

$$\frac{\partial \tau_{23}}{\partial x_3} = \frac{\partial P}{\partial x_2}, \quad \tau_{23} = \eta \frac{\partial u_2}{\partial x_3}, \tag{2.120}$$

$$\frac{\partial P}{\partial x_3} = 0, \tag{2.121}$$

with velocity boundary conditions

$$u_1 = U_1 \text{ and } u_2 = U_2 \text{ at } x_3 = h, \tag{2.122}$$

$$u_1 = u_2 = 0 \text{ at } x_3 = 0. \tag{2.123}$$

Integration of Eqs. 2.119 and 2.120 with respect to x_3 gives

$$u_1 = U_1 \frac{x_3}{h} - \frac{x_3}{2\eta} \frac{\partial P}{\partial x_1} (h - x_3), \tag{2.124}$$

$$u_2 = U_2 \frac{x_3}{h} - \frac{x_3}{2\eta} \frac{\partial P}{\partial x_2} (h - x_3), \tag{2.125}$$

and, by continuity,

$$\frac{\partial}{\partial x_1} (\bar{u}_1 h) + \frac{\partial}{\partial x_2} (\bar{u}_2 h) + \frac{\partial h}{\partial t} = 0, \tag{2.126}$$

where the gap-wise averaged velocity components $\bar{u}_1$ and $\bar{u}_2$ are given by, respectively,

$$\bar{u}_1 = \frac{1}{h} \int_0^h u_1 dx_3 = \frac{1}{2} U_1 - \frac{h^2}{12\eta} \frac{\partial P}{\partial x_1}, \tag{2.127}$$

$$\bar{u}_2 = \frac{1}{h} \int_0^h u_2 dx_3 = \frac{1}{2} U_2 - \frac{h^2}{12\eta} \frac{\partial P}{\partial x_2}. \tag{2.128}$$

Substituting Eqs. 2.115 and 2.116 into 2.114 yields

$$\frac{\partial}{\partial x_1} \left(\frac{h^3}{\eta} \frac{\partial P}{\partial x_1} \right) + \frac{\partial}{\partial x_2} \left(\frac{h^3}{\eta} \frac{\partial P}{\partial x_2} \right) = 6U_1 \frac{\partial h}{\partial x_1} + 6U_2 \frac{\partial h}{\partial x_2} + 12 \frac{\partial h}{\partial t}, \tag{2.129}$$

which is a fundamental equation of lubrication theory known as the Reynolds equation. Neglect of compressibility is implied in Eq. 2.114. Dowson (1962) has given a more general formulation that permits variation in the density of the fluid. Lubrication theory for non-Newtonian fluids including viscoelastic fluids has been discussed in detail by Tanner (1960, 2000). In the next chapter, we will see that the lubrication simplification is useful for injection molding flow analysis.

Chapter 3
Mold Filling and Post Filling

3.1 Hele-Shaw Equation

3.1.1 Flow in Thin Cavity of Arbitrary In-plane Dimensions

Most injection-molded parts are thin walled, i.e., they have a small thickness compared to other typical dimensions. Therefore, one can reduce the three-dimensional flow to a simpler two-dimensional problem, using the lubrication approximation (Richardson 1972). We consider a polymer flow through a thin cavity with a slowly varying gap-wise dimension and arbitrary in-plane dimensions. Assume that x_1, x_2 are the planar coordinates, x_3 is the gap-wise direction coordinate. The flow occurs between two walls at $x_3 = \pm h/2$. Adjacent to each wall there is a frozen layer of the solidified polymer so that the polymer melt flows between two solid–liquid interfaces at $x_3 = s^-(x_1, x_2)$ and $x_3 = s^+(x_1, x_2)$ (see Fig. 3.1).

The equations of motion are then given by

$$\frac{\partial P}{\partial x_i} = \frac{\partial \tau_{i3}}{\partial x_3}, \quad (i = 1,\ 2), \tag{3.1}$$

where τ_{i3} is the shear stress components τ_{13} and τ_{23}. Integration of (3.1) with respect to x_3 leads to

$$\tau_{i3} = x_3 \frac{\partial P}{\partial x_i} + B_i, \quad (i = 1,\ 2), \tag{3.2}$$

where B_i is a vector

$$B_i = \tau_{i3}(s^-) - s^- \frac{\partial P}{\partial x_i}, \quad (i = 1,\ 2). \tag{3.3}$$

It is convenient to use a fluidity Φ here (Huilgol 2006), which provides a way of defining the shear rate in terms of the shear stress, such that

R. Zheng et al., *Injection Molding*, DOI: 10.1007/978-3-642-21263-5_3,

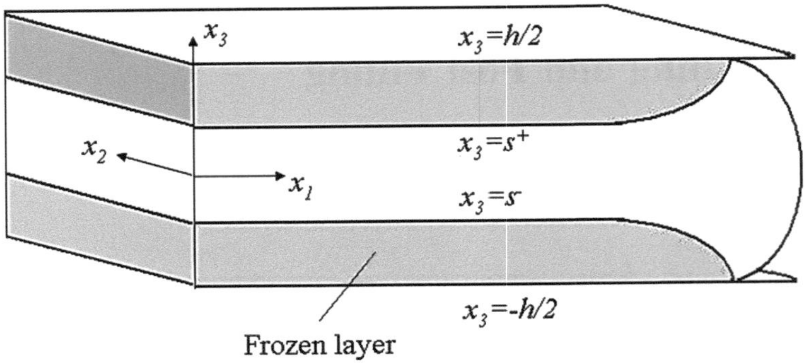

Fig. 3.1 Definition of the local coordinate system for mold cavity

$$\frac{\partial u_i}{\partial x_3} = \Phi \tau_{i3}, \quad (i = 1,\ 2), \tag{3.4}$$

where u_i is the velocity vector with non-zero components u_1 and u_2. Then we obtain, from (3.2)

$$\frac{\partial u_i}{\partial x_3} = \Phi x_3 \frac{\partial P}{\partial x_i} + \Phi B_i, \quad (i = 1,\ 2). \tag{3.5}$$

Thus, using the no-slip condition at $x_3 = s^-$, we obtain the velocity field

$$u_i(x_3) = \frac{\partial P}{\partial x_i}\left(\int\limits_{s^-}^{x_3} \Phi z dz\right) + B_i \int\limits_{s^-}^{x_3} \Phi dz, \quad (i = 1,\ 2). \tag{3.6}$$

Since u_i (s^+) = 0, we have

$$\frac{\partial P}{\partial x_i}\left(\int\limits_{s^-}^{s^+} \Phi z dz\right) + B_i \int\limits_{s^-}^{s^+} \Phi dz = 0, \quad (i = 1,\ 2). \tag{3.7}$$

Hence

$$B_i = -\frac{\int_{s^-}^{s^+} \Phi z dz}{\int_{s^-}^{s^+} \Phi dz}\left(\frac{\partial P}{\partial x_i}\right), \quad (i = 1,\ 2). \tag{3.8}$$

Let

$$C = \frac{\int_{s^-}^{s^+} \Phi z dz}{\int_{s^-}^{s^+} \Phi dz}, \tag{3.9}$$

then the velocity field can be rewritten as

$$u_i = \left(\int_{s^-}^{x_3} \Phi z dz - C \int_{s^-}^{x_3} \Phi dz \right) \frac{\partial P}{\partial x_i}, \quad (i = 1,\ 2). \tag{3.10}$$

We define the gap-wise average velocity field $\bar{u}_i$ through

$$\bar{u}_i = \frac{1}{s^+ - s^-} \int_{s^-}^{s^+} u_i dx_3, \quad (i = 1,\ 2). \tag{3.11}$$

Thus we have

$$\bar{u}_i = \frac{1}{s^+ - s^-} \left[\int_{s^-}^{s^+} \left(\int_{s^-}^{x_3} \Phi z dz \right) dx_3 - C \int_{s^-}^{s^+} \left(\int_{s^-}^{x_3} \Phi dz \right) dx_3 \right] \frac{\partial P}{\partial x_i}, \quad (i = 1,\ 2). \tag{3.12}$$

Integration by parts gives

$$\int_{s^-}^{s^+} \left(\int_{s^-}^{x_3} \Phi z dz \right) dx_3 = s^+ \int_{s^-}^{s^+} \Phi x_3 dx_3 - \int_{s^-}^{s^+} \Phi x_3^2 dx_3, \tag{3.13}$$

$$\int_{s^-}^{s^+} \left(\int_{s^-}^{x_3} \Phi dz \right) dx_3 = s^+ \int_{s^-}^{s^+} \Phi dx_3 - \int_{s^-}^{s^+} \Phi x_3 dx_3. \tag{3.14}$$

Substituting Eqs. 3.13 and 3.14 into 3.12 and replacing C by its form in (3.9) gives

$$\bar{u}_i = \frac{1}{s^+ - s^-} \left[- \int_{s^-}^{s^+} \Phi x_3^2 \, dx_3 + \frac{\left(\int_{s^-}^{s^+} \Phi x_3 dx_3 \right)^2}{\int_{s^-}^{s^+} \Phi dx_3} \right] \frac{\partial P}{\partial x_i}, \quad (i = 1,\ 2), \tag{3.15}$$

which can be rewritten as

$$\bar{u}_i = - \frac{S}{s^+ - s^-} \left(\frac{\partial P}{\partial x_i} \right), \quad (i = 1,\ 2), \tag{3.16}$$

where

$$S = \int_{s^-}^{s^+} \Phi x_3^2 dx_3 - \frac{\left(\int_{s^-}^{s^+} \Phi x_3 dx_3 \right)^2}{\int_{s^-}^{s^+} \Phi dx_3}. \tag{3.17}$$

This result applies to fluids for which fluidity appears in a viscometric flow. For generalized Newtonian fluids, $\Phi = 1/\eta$, the equation for S is the same as that in the book of Kennedy (1995).

Recall that the mass conservation equation is

$$\frac{\partial u_i}{\partial x_i} = -\frac{1}{\rho}\frac{\mathrm{D}\rho}{\mathrm{D}t}, \quad (i = 1, 2, 3), \tag{3.18}$$

where ρ is the density of the fluid. Integration of the equation of mass conservation with respect to the gap-wise coordinate x_3 from $-h/2$ to $h/2$ leads to

$$\frac{\partial}{\partial x_i}[(s^+ - s^-)\bar{u}_i] + \int_{-h/2}^{h/2} \frac{\partial u_3}{\partial x_3} dx_3 = -\int_{-h/2}^{h/2} \frac{1}{\rho}\frac{\mathrm{D}\rho}{\mathrm{D}t} dx_3, \quad (i = 1, 2). \tag{3.19}$$

Noting that

$$\int_{-h/2}^{h/2} \frac{\partial u_3}{\partial x_3} dx_3 = \frac{\partial h}{\partial t}, \tag{3.20}$$

by substituting (3.16) and (3.20) into (3.19), we obtain the equation

$$\frac{\partial}{\partial x_i}\left(S\frac{\partial P}{\partial x_i}\right) = \int_{-h/2}^{h/2} \frac{1}{\rho}\frac{\mathrm{D}\rho}{\mathrm{D}t} dx_3 + \frac{\partial h}{\partial t}, \quad (i = 1,\ 2)\,. \tag{3.21}$$

This resulting equation is known as the Hele-Shaw equation.

The term $\partial h/\partial t$ is included to account for the mold deformation, or to allow the model to be applied to compression molding. One has $\partial h/\partial t = 0$ when the walls are stationary.

During the filling stage of the injection molding, the density variations are negligible for temperatures and pressures of interest. Hence, the first term on the right-hand-side of (3.21) can be eliminated. That is

$$\int_{-h/2}^{h/2} \frac{1}{\rho}\frac{\mathrm{D}\rho}{\mathrm{D}t} dx_3 = 0. \tag{3.22}$$

During packing and cooling stages, the dependence of the density on pressure and temperature must be considered, but the convective term can be neglected, so that the first term on the right-hand-side of (3.21) can be decomposed as

$$\int_{-h/2}^{h/2} \frac{1}{\rho}\frac{\mathrm{D}\rho}{\mathrm{D}t} dx_3 = \left[\int_{-h/2}^{h/2} \frac{1}{\rho}\left(\frac{\partial\rho}{\partial P}\right)_T dx_3\right]\frac{\partial P}{\partial t} + \int_{-h/2}^{h/2} \left[\frac{1}{\rho}\left(\frac{\partial\rho}{\partial T}\right)_P \frac{\partial T}{\partial t}\right] dx_3. \tag{3.23}$$

For semi-crystalline materials, the density is a function of the relative crystallinity α (see Chap. 4), and therefore an extra term appears on the right-hand side of the above equation: $\int_{-h/2}^{h/2}(1/\rho)(\partial\rho/\partial\alpha)(D\alpha/Dt)dx_3$.

For a geometrically complex part, some regions may be filled first while others are not. The filled regions will experience a packing type flow, and the unfilled regions are still experiencing filling. Using the unified equation (3.21), the situation can be handled automatically.

We now define:

$$\beta = -\frac{1}{\rho}\left(\frac{\partial\rho}{\partial T}\right)_P, \kappa = \frac{1}{\rho}\left(\frac{\partial\rho}{\partial P}\right)_T, \text{ and } \bar{\kappa} = \frac{1}{h}\int_{-h/2}^{h/2}\kappa dx_3, \tag{3.24}$$

where β and k are the thermal expansion coefficient and the isothermal compressibility respectively, and $\bar{\kappa}$ is the gap-wise average value of compressibility.

After rearrangement, Eq. (3.21) (with $\partial h/\partial t = 0$) can be rewritten in the form

$$\frac{\partial}{\partial x_i}\left(S\frac{\partial P}{\partial x_i}\right) = K_P\frac{\partial P}{\partial t} + \Pi_T + H_\alpha \quad (i = 1, 2), \tag{3.25}$$

where

$$K_P = \bar{\kappa}h, \tag{3.26}$$

$$\Pi_T = -\int_{-h/2}^{h/2}\beta\frac{\mathrm{D}T}{\mathrm{D}t}dx_3, \tag{3.27}$$

$H_\alpha = 0$ for amorphous materials, and

$$H_\alpha = \int_{-h/2}^{h/2}\frac{1}{\rho}\frac{\partial\rho}{\partial\alpha}\frac{\mathrm{D}\alpha}{\mathrm{D}t}dx_3 \tag{3.28}$$

for semi-crystalline materials. The value of $\partial\rho/\partial\alpha$ is approximately $\rho_s - \rho_m$ where ρ_m and ρ_s stand for the melt density and the solid density, respectively. Chiang et al. (1991) consider a possible discontinuity in the density at the melt-solid transition due to crystallization and express the H_α term as

$$H_\alpha = (\ln\rho_s - \ln\rho_m)_{x_3=s^-}\frac{\partial s^-}{\partial t} + (\ln\rho_m - \ln\rho_s)_{x_3=s^+}\frac{\partial s^+}{\partial t}, \tag{3.29}$$

We have the following boundary conditions in the gap-wise direction:

$$u_1 = u_2 = u_3 = 0 \text{ at } x_3 = s^+ \text{ and } x_3 = s-. \tag{3.30}$$

Further, in the planar coordinates, an impermeability condition (zero normal velocity) is applied along the edges of the cavity, and along the edges of inserts if there are any. This condition is equivalent to a natural boundary condition for pressure, i.e., $\partial P/\partial n = 0$, where $\partial/\partial n$ denotes the gradient in the direction normal to the boundary. During the filling stage, the value of the pressure at the advancing flow front is taken to be zero. Finally, during the filling stage, a specified total flow rate is imposed at the inlet boundary. During the post-filling stage, an inlet pressure profile is specified, which is assumed to be uniform along the cross section, but time dependent.

In injection molding the pressure level can be very high, so that the assumption of pressure-independent viscosity can lead to incorrect simulation results. Some authors (e.g., Sherbelis and Friedl 1996) have considered pressure-dependent viscosity in their simulations. However, caution should be made because the pressure-dependent viscosity might invalidate the Hele-Shaw approximation. When the viscosity varies with pressure, the pressure gradient in the flow direction (say, x_1 direction) and that in the thickness direction have the following relation: $\partial P/\partial x_3 = \dot{\gamma}_{13}(\partial\eta/\partial P)(\partial P/\partial x_1)$ (Huilgol and You 2006). Only if $\dot{\gamma}_{13}(\partial\eta/\partial P) \ll 1$, can the dependence of pressure on x_3 be neglected and the Hele-Shaw equation be used. See, also, Dowson (1962).

The Hele-Shaw equation solves for the pressure problem, which is coupled with the temperature equation:

$$\rho c_p \frac{\mathrm{D}T}{\mathrm{D}t} = \frac{\partial}{\partial x_i}\left(k_{ij}\frac{\partial T}{\partial x_j}\right) + \tau_{ij}D_{ij} + \frac{T}{\rho}\left(\frac{\partial \rho}{\partial T}\right)_P \frac{\mathrm{D}P}{\mathrm{D}t} + \dot{Q}, \tag{3.31}$$

where c_p and k_{ij} are the specific heat and the thermal conductivity, respectively. The thermal conductivity takes the tensor form to account for anisotropy (Huilgol et al. 1992).

For thin cavity flows, the in-plane thermal conduction can be neglected and hence the temperature equation reduces to

$$\rho c_p \frac{\mathrm{D}T}{\mathrm{D}t} = \frac{\partial}{\partial x_i}\left(k_{33}\frac{\partial T}{\partial x_3}\right) + \tau_{ij}D_{ij} + \frac{T}{\rho}\left(\frac{\partial \rho}{\partial T}\right)_P \frac{\mathrm{D}P}{\mathrm{D}t} + \dot{Q}. \tag{3.32}$$

The thermal boundary conditions in the gap-wise direction are:

$$-k\frac{\partial T}{\partial x_3} = \tilde{h}(T - T_w^-) \text{ at } x_3 = -h/2, \tag{3.33}$$

and

$$-k\frac{\partial T}{\partial x_3} = \tilde{h}(T - T_w^+) \text{ at } x_3 = h/2. \tag{3.34}$$

Here $-k(\partial T/\partial x_3)$ is the heat flux, $\tilde{h}$ is the heat transfer coefficient (Delaunay et al. 2000a), T_w^- and T_w^+ are the mold wall temperatures at the two opposite sides of the cavity. The mold wall temperature should be calculated from the mold cooling analysis (see Chap. 7).

No thermal boundary conditions are required along the edge boundaries in the x_1–x_2 plane due to the neglect of the thermal conduction in the x_1–x_2 plane. The inlet temperature is assumed uniform and taken as the injection melt temperature. The flow-front temperature boundary condition requires a special treatment to mimic the fountain flow effect, which will be discussed later in Chap. 8.

3.1.2 Axisymmetric Flow in a Tube

Similarly, the flow in a runner can be modeled as an axisymmetric one-dimensional flow in a tube of radius R. For a stationary wall, we have the pressure equation as

$$\frac{\partial}{\partial x}\left(S_{\mathrm{cyl}}\frac{\partial P}{\partial x}\right)=\int_0^R \frac{1}{\rho}\frac{\mathrm{D}\rho}{\mathrm{D}t}rdr, \tag{3.35}$$

where x is the axial coordinate, r is the radial coordinate, and

$$S_{\mathrm{cyl}}=\frac{1}{4}\int_0^{r^+}\Phi r^3 dr, \tag{3.36}$$

in which r^+ is the solid–melt interface location.

The temperature equation in the cylindrical coordinates (x, r, θ) is

$$\rho c_p\left(\frac{\partial T}{\partial t}+u_x\frac{\partial T}{\partial x}\right)=\frac{1}{r}\frac{\partial}{\partial r}\left(rk_r\frac{\partial T}{\partial r}\right)+\tau_{rx}\frac{\partial u_x}{\partial r}+\frac{T}{\rho}\left(\frac{\partial\rho}{\partial T}\right)_P\frac{\mathrm{D}P}{\mathrm{D}t}+\dot{Q}, \tag{3.37}$$

where u_x is the axial flow velocity, and k_r is the thermal conductivity in the radial direction. In this equation, convection in the radial direction and conduction in the axial direction have been neglected.

3.2 Frozen Layer

In the above section, we made use of the concept of a “frozen layer”. The thickness of the frozen layer is very important to flow analysis since it reduces the effective melt channel. The narrower effective melt channel results in increased flow resistance. This means that, if the frozen layer thickness or the location of the solid–liquid interface is not accurately calculated, the injection pressure may be inaccurately estimated. Janeschitz-Kriegl (1977, 1979) has considered an analytical solution of the growth of the frozen layer for an idealized mold filling problem, where the polymer melt flows steadily through an infinitely long duct of flat rectangular cross-section. Now, using numerical methods we can deal with

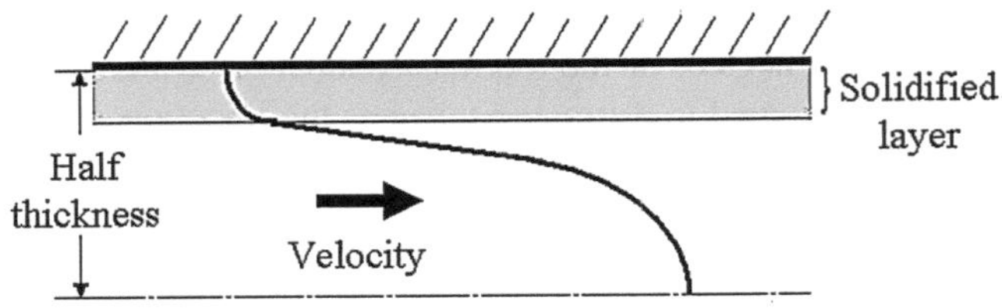

Fig. 3.2 A sketch of viscosity distribution across the cavity thickness

more complex geometries. What is still contentious is the solidification criterion. There are different ways in which the thickness of the frozen layer or the location of the solid–liquid interface can be determined:

1. Calculate the gap-wise velocity distribution. The thickness of frozen layer is determined by the effective zero velocity region (Fig. 3.2). This calculation requires an accurate prediction of viscosity ranging from the processing temperature down to the temperature of the material at the mold wall.
2. Use the concept of a "no-flow temperature" which indicates the temperature at which the stagnation of flow occurs (see, for example, Janeschitz-Kriegl 1979; Kennedy 1995). Although most reported studies simply assume a constant value of the no-flow temperature, the no-flow temperature is related to processing. This is especially true for semicrystalline materials. It would be more appropriate to use a no-flow temperature depending on crystallization.
3. Use the increase of viscosity relative to the viscosity at the initial melt temperature as a no-flow condition. Janeschitz-Kriegl (1977) defines an increase of viscosity by a factor of 2.7 as a criterion of solidification.

Both methods 1 and 3 involve using of constitutive model for viscosity prediction. Care should be taken to ensure the realistic behavior of the viscosity model chosen. Kennedy (1995) has shown that, while some existing models can deal with fluid behavior above the melting point, the extrapolation to low temperatures does not predict the sharp increase in viscosity for the semicrystalline polymer. To produce a sudden rise in viscosity, we may need to incorporate the crystallization kinetics in rheology. This will be discussed in more detail in the next chapter.

3.3 Mold Deformation

A common hypothesis in injection molding simulation is to assume an infinitely rigid mold. However, in reality, the high cavity pressure can lead to mold deflection due to mold compliance. Experimental investigations (Leo and Cuvelliez 1996; Delaunay et al. 2000b) have shown that the mold elastic deformation can play a significant role in the pressure–time history.

Recall Eq. 3.21 in Sect. 3.1.1, where the term $\partial h/\partial t$ is set to zero for rigid walls. This term should be included if we want to consider the effect of mold deflection. We write it as

$$\frac{\partial h}{\partial t} = \frac{\partial h}{\partial P}\frac{\partial P}{\partial t}. \qquad (3.38)$$

In the case of mold deformation, the cavity thickness h is a function of the local pressure. Baaijens (1991) simply chose $\partial h/\partial P = 0.4$ μm/MPa and showed that a small amount of mold compliance can have a significant influence on the cavity pressure history, and neglect of the mold elasticity will lead to under-prediction of the cavity pressure. Pantani et al. (2001) assume the following relationship:

$$h(x, P(x,t)) = h_0(x)[1 + C_M P(x,t)], \qquad (3.39)$$

where $h_0(\boldsymbol{x})$ is the initial cavity thickness before deflection, C_M has the dimension of compliance. Typical values of C_M range from 10^{-4}–10^{-3} MPa^{-1}, depending on mold geometry and material. Differentiating Eq. 3.39 with respect to P gives

$$\frac{\partial h}{\partial P} = h_0 C_M, \qquad (3.40)$$

and hence (3.38) becomes

$$\frac{\partial h}{\partial t} = h_0 C_M \frac{\partial P}{\partial t}. \qquad (3.41)$$

Thus, with the incorporation of Eq. 3.41, one obtains, instead of Eqs. 3.25–3.27:

$$\frac{\partial}{\partial x_i}\left(S\frac{\partial P}{\partial x_i}\right) = K_P\frac{\partial P}{\partial t} + \Pi_T + H_\alpha \quad (i = 1,\, 2), \qquad (3.42)$$

where

$$K_P = [\bar{\kappa} + C_M(1 + \bar{\kappa}P)]h_0, \qquad (3.43)$$

$$\Pi_T = -\int_{-h(x,t)/2}^{h(x,t)/2} \beta\frac{\mathrm{D}T}{\mathrm{D}t}dx_3, \qquad (3.44)$$

$H_\alpha = 0$ for amorphous materials, and

$$H_\alpha = \int_{-h(x,t)/2}^{h(x,t)/2} \frac{1}{\rho}\frac{\partial\rho}{\partial\alpha}\frac{\mathrm{D}\alpha}{\mathrm{D}t}dx_3, \qquad (3.45)$$

for semi-crystalline materials.

Equation 3.43 indicates that the effect of mold deformation corresponds to an increase in material compressibility.

In the post-filling stage, when the gate freezes, the left-hand-side term in (3.42) is negligible, we then have

$$\frac{\partial P}{\partial t} = -\frac{\Pi_T + H_\alpha}{[\bar{\kappa} + C_M(1 + \bar{\kappa}P)]h_0}, \tag{3.46}$$

which shows that the cooling-induced contraction is the driving force for pressure decay, while the effect of the mold deformation tends to slow down the pressure decay.

A problem known as the *core shift* is closely related to the mold deformation problem. A core is the part of a mold that shapes the inside of a molded product. Core shift is the spatial deviation of the position of the core caused by non-uniform pressure distribution over the surface of the core during the filling and packing stages. Prediction of the core shift in injection molding has been attempted by some researchers, e.g., Bakharev et al. (2004).

3.4 Wall Slip

In the derivation of the Hele-Shaw equation, we have applied the so-called "no-slip" boundary condition at the liquid–solid interface, where the liquid is assumed to adhere, thus to have no velocity relative to the solid surface.

This assumption holds very well for Newtonian liquids, but it does not seem to be generally valid for flows of polymeric liquids. Extrusion experiments with polymer melts have shown that slip over solid surfaces may appear when the wall shear stress exceeds a critical value (usually of the order 0.1 MPa), see, for example, Ramamurthy (1986), Kalika and Denn (1987), Hatzikiriakos and Dealy (1991, 1992) and Hatzikiriakos (1993). The appearance of distortions on the extrudate surface is believed to be a consequence of wall slip. Lim and Schowalter (1989) have distinguished four flow regimes in the extrusion of a narrow molecular weight polybutadienes in a slit die. The first regime is characterized by a smooth, glossy extrudate and an almost Newtonian flow behavior. Loss of gloss at a critical wall shear stress about 0.1 MPa indicates the start of the second flow regime. The third flow regime starts at a wall shear stress of the order 0.2 MPa, where the shear stress becomes independent of shear rate, and pressure oscillations set in, indicating a stick–slip flow pattern. In the fourth flow regime, the pressure oscillation amplitude and frequency are reduced, and the flow is nearly steady with a non-zero velocity at the liquid–solid interface.

Wall slipping has also been proposed as a possible mechanism for some surface defects of injection-molded parts (Denn 2001). In a study of several blends of BPA polycarbonate and ABS resins, Hobbs (1996) found that periodic surface irregularities can be made at high injection rates. The periodicity is produced due to the oscillation of the flow front between the two mold walls, as the wall slip first occurs on one surface and then the other.

Early microstructural discussions suggested that slip occurs because the polymer molecules at the wall align themselves more strongly with the flow than those away from the wall. Pearson and Petrie (1968) consider that the ratio of molecule size to surface roughness scale is important: when the molecular size is smaller than the wall roughness scale, then no effective slip can occur, but when large macromolecules are present, slip may well happen. Jabbarzadeh et al. (1999, 2000) used a molecular dynamics simulation to study the effect of the wall roughness on the wall slip, where the wall was modeled by a simple sinusoidal wall. Results showed that the wall slip increases with the wall roughness period and the size of molecules, while it decreases with increasing wall roughness amplitude. Their findings confirm the theory of Pearson and Petrie.

Some phenomenological approaches for handling the slip boundary condition have been proposed, where the slip velocity u_s is takes as an empirical function of the wall shear stress τ_w. Rosenbaum and Hatzikiriakos (1997) introduced a power-law relation for slip velocity:

$$u_s = \frac{a}{1 + \left(\tau_c / \tau_w\right)^{10}} \tau_w^n, \tag{3.47}$$

where τ_c is the critical wall stress for slip, a is a scalar coefficient and n is a power-law index. The factor $1/\left[1 + (\tau_c/\tau_w)^{10}\right]$ is effectively zero for wall stresses less than τ_c and approaches one when τ_w exceeds τ_c.

Phan-Thien (1988) and Tanner (1994) have considered the situation of non-uniform stresses at outlet and inlet. Thus the idea and methods regarding the partial slip were discussed.

So far, injection molding analyses involving wall slip have been scarce, although it is clear that one cannot ignore the possible slip when high stresses ($\tau_w \geq 0.1$ MPa) exist. This is especially important for micro-injection moldings [see, for example, Coates (2007) and Zhang et al. (2008)]. Phenomenological models can be a good starting point for the simulation of the wall slip. However, it is almost certain that, to make real progress in this area, one needs to consider the microstructure of the fluid–solid interface.

Chapter 4
Crystallization

4.1 Introduction

The majority of polymers used in industry are semicrystalline. Crystallization occurs during the processing, and consists of two stages:

1. *Nucleation*: Active nuclei are formed within the liquid phase, from which crystals can develop. If the nuclei appear spontaneously due to thermal fluctuations in the liquid, the nucleation is called *homogeneous nucleation*. If the nuclei are formed on the surface of foreign substances or crystals of the same material already present in the melt, the nucleation is called *heterogeneous nucleation*.
2. *Growth*: The nuclei develop into observable crystals.

In the formation of crystals, polymer chains fold back and forth to form the crystalline lamellae. The crystalline lamellae and the amorphous phase are arranged in semicrystalline morphological entities, ranging from a micron to several millimeters in size. The most common morphologies that can be found in injection-molded polymers are spherulites, which usually form under quiescent conditions, and shish-kebab structures, which may appear under shear flow [see, for example, Eder and Janeschitz-Kriegl (1997), Zuidema et al. (2001) and Janeschitz-Kriegl (2009)].

In the filling stage of injection molding, the polymer melt is subjected to high shear rates, which results in the so-called flow-induced crystallization (FIC). The main features of FIC are the dramatic enhancement of the crystallization rate, and the formation of oriented morphology. The effect of FIC continues after the cessation of flow.

Final properties of an injection-molded product, made of semi-crystalline polymers, strongly depend on the flow-induced microstructure (crystallinity, morphology, orientation, etc.). Particularly, the enhancement in polymer crystallization rate crucially changes the solidification behavior, and the oriented microstructure leads to a local anisotropy in thermal and mechanical properties,

R. Zheng et al., *Injection Molding*, DOI: 10.1007/978-3-642-21263-5_4,

which can further cause anisotropic shrinkage and hence enhance warpage of the injection-molded parts.

In this chapter, we shall focus on two relevant topics: the effect of flow on crystallization kinetics, and the effect of crystallization on rheological and thermal properties.

4.2 Crystallization Kinetics

4.2.1 The Kolmogoroff-Avrami Model

We model the effect of flow on crystallization kinetics by considering growth and nucleation. Following the work of Kolmogoroff (1937), we first consider an unrestricted growth of crystals and calculate a fictive volume fraction. Let G denote the rate of growth of the spherulite radius as a function of time. The volume of a phantom spherulite, nucleated at the instant s and grown up to the instant t, is given by $\upsilon_1(s,t) = (4\pi/3)\left[\int_s^t G(u)du\right]^3$; the expression in the square brackets is the radius of the spherulite. The shish-kebab structure is represented by a cylindrical geometry. It is assumed by Eder and Janeschitz-Kriegl (1997) that the lateral growth rate is the same as the spherulite radial growth rate G, unaffected by flow. Its volume is $\upsilon_2(s,t) = \pi l_s\left[\int_s^t G(u)du\right]^2$, where l_s is the length of the shish at time t. Denoting the number density of the quiescent nuclei by N_q and the flow-induced nuclei number density by N_f, the unrestricted total volume fraction of phantom crystals at time t is then given by

$$\varphi(t) = \int_0^t \dot{N}_q(s)\upsilon_1(s,t)ds + \int_0^t \dot{N}_f(s)[\omega\upsilon_1(s,t) + (1-\omega)\upsilon_2(s,t)]ds. \qquad (4.1)$$

Here the first integral term on the right-hand side of the equation is the quiescent contribution. The second integral term is the contribution of the flow-induced crystallization, where we introduce a two-value weight function ω:

$$\omega = \begin{cases} 0 & \text{if } \dot{\gamma} > 1/\lambda_R, \text{ and } \int_0^{t_s} \eta\dot{\gamma}^2 dt > w_c \\ 1 & \text{otherwise} \end{cases}, \qquad (4.2)$$

where λ_R is the longest Rouse time, η is the shear viscosity, $\dot{\gamma}$ is the generalized shear rate defined as $\dot{\gamma} = \sqrt{2D_{ij}D_{ij}}$, with $D_{ij} = (1/2)(\partial u_i/\partial x_j + \partial u_j/\partial x_i)$ being the rate-of-deformation tensor, t_s is the shearing time, the integration gives the specific work done on the sheared polymer, and w_c is the critical specific work. It has been noticed by van Meerveld et al. (2004) that there is a minimum shear rate $\dot{\gamma}_{\min} \sim 1/\lambda_R$ below which the orientated structure (shish-kebab) is unlikely to be formed. Janeschitz-Kriegl et al. (2003) suggests that it is the specific work done on the sheared polymer melt that controls the resulting morphology. Experimental observations of Mykhaylyk et al. (2008) support this theory. They found that, above $\dot{\gamma} = 1/\lambda_R$, a critical amount of work is necessary to cause a transition from isotropic spherulites to oriented shish-kebabs.

In order to account for impingement, the fictive volume fraction should be converted to the actual relative crystallinity, α, which is also called the degree of space filling, defined as the ratio of crystallized volume at a given time to the total crystallizable volume. According to Kolmogoroff (1937) and Avrami (1939), the actual crystal volume increase is equal to the liquid fraction of the fictitious increase, i.e., $d\alpha = (1 - \alpha)d\varphi$, which can be integrated to give

$$\alpha = 1 - \exp(-\varphi). \tag{4.3}$$

This means that, to calculate the relative crystallinity, we have to calculate only the fictive volume fraction φ. But to obtain φ, we need to consider the growth rate and the nuclei number density.

4.2.2 *Growth Rate*

The increasing number of activated nuclei is the first noticeable effect induced by flow. The effect of flow on growth rate was also observed (Monasse 1995), but it is usually less important and can be neglected (Koscher and Fulchiron 2002), while the effect of flow on nucleation must be considered.

For a spherulite, the radial growth rate G can be calculated by applying the Hoffman–Lauritzen theory (Lauritzen and Hoffman 1960)

$$G(T) = G_0 \exp\left[-\frac{U^*}{R_g(T - T_\infty)}\right] \exp\left[-\frac{K_g(T + T_m^0)}{2T^2 \Delta T}\right], \tag{4.4}$$

where U^* is the activation energy of motion, often taken as a generic value 6,270 J/mol (e.g., Koscher and Fulchiron 2002), R_g is the gas constant, $T_\infty = T_g - 30$ with T_g being the glass transition temperature, $\Delta T = T_m^0 - T$ is the degree of supercooling, where T_m^0 is the equilibrium melting temperature and T is the temperature at the crystallization. All the temperatures in this equation are given in K. The value of T_m^0 under atmospheric pressure can be determined by the Hoffman–Weeks extrapolation method (Hoffman and Weeks 1962). In this method, the melting temperature T_m, measured using the Differential Scanning Calorimetry (DSC), is plotted as a function of the crystallization temperature, T_c, and the $T_m = f(T_c)$ curve is extrapolated up to its intersection with the $T_m = T_c$ straight line. The intersection gives the value of T_m^0 (Fig. 4.1). The Hoffman–Weeks extrapolation method, however, is not a very accurate method. Improved experimental methods have also been discussed by Marand et al. (1998) and Al-Hussein and Strobl (2002).

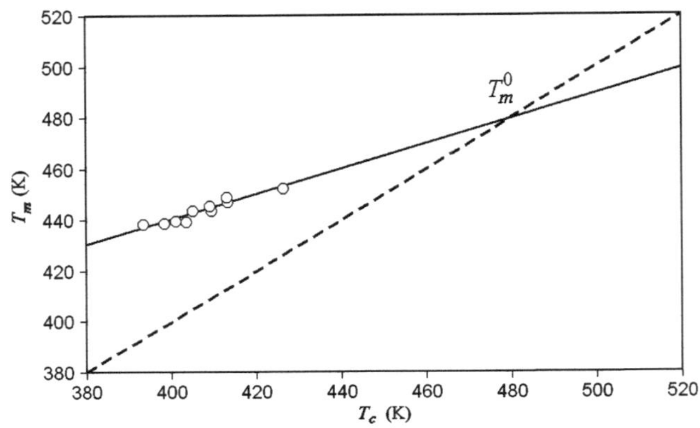

Fig. 4.1 Determination of equilibrium melting temperature for a sample of iPP by the Hoffman–Weeks extrapolation method (From Zheng et al. (2010), with permission from Springer Science + Business Media B.V.)

The equilibrium melting temperature may depend on pressure. Fulchiron et al. (2001) express the pressure dependence as a polynomial function

$$T_m^0(P) = T_m^0(0) + a_1 P + a_2 P^2, \tag{4.5}$$

where P is the pressure, and a_1 and a_2 are constants that can be determined from the Pressure–Volume–Temperature diagram.

To determine the parameters G_0 and K_g, one needs to measure the growth rate $G(T)$. For materials with slow crystallization kinetics, one can easily measure the spherulite growth rate as a function of temperature from micrographs (Fig. 4.2). Then G_0 and K_g are determined by plotting $\ln G + U^*/R_g(T - T_\infty)$ against $(T + T_m^0)/2T^2\Delta T$.

For some industrial polymers, crystallization rates are too high so that the observation of spherulite growth in the interesting temperature range is not experimentally possible. An approximate method is therefore useful. Van Krevelen (1976) proposed a semi-empirical equation for the growth rate of variety of common polymers as follows:

$$\log G = \log G_0 - 2.3\frac{T_m^0}{T}\left(\frac{T_m^0}{T_m^0 - T_g} + \frac{50}{T_m^0 - T}\right), \tag{4.6}$$

with $G_0 = 7.5 \times 10^8$ μm/s

For shish-kebab structures, one assumes that the lateral growth rate is equal to the corresponding radial growth rate for spherulites, but it is not yet possible to determine the length growth rate. A useful approach to deal with the growth of thread-like nuclei has been proposed by Liedauer et al. (1993). The method involves the total length of threads per unit volume, which is given by

$$L_{tot} = \int_0^t \dot{N}_f(s) l_s(t - s) ds. \tag{4.7}$$

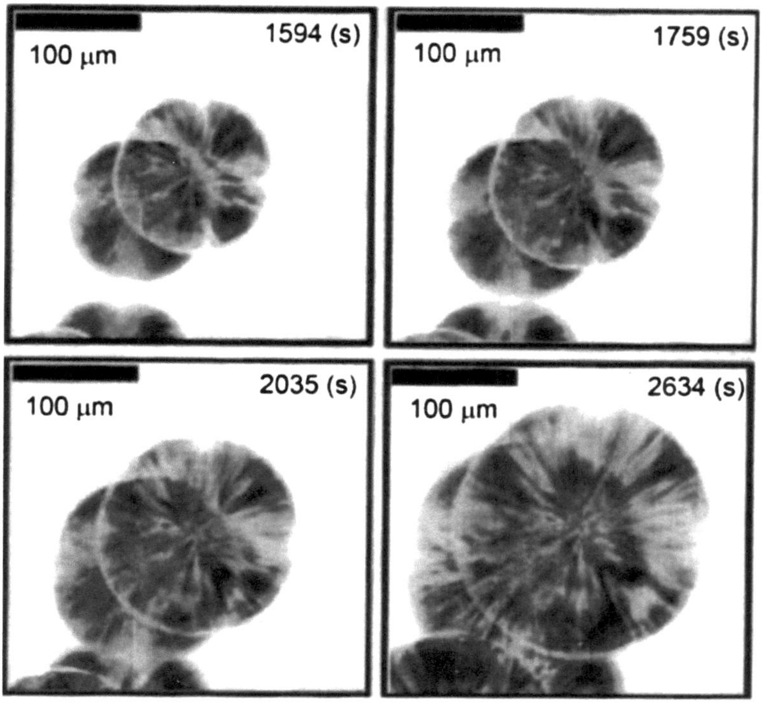

Fig. 4.2 Growth of spherulites observed under microscopy for iPP at 135°C (From Zheng et al. (2010), with permission from Springer Science + Business Media B.V.)

The total length per unit volume has the following relationship with the average distance D between the threads:

$$L_{\text{tot}} = \frac{2}{\sqrt{3}D^2}, \tag{4.8}$$

The distance D is possible to be measured from electron micrographs, or predicted from birefringence experimental data. The function of $l_s(t)$ is assumed to be

$$l_s(t) = g_l(\lambda_R \dot{\gamma})^2 t, \tag{4.9}$$

where g_l is a constant with the dimension $\text{m}^{-2}\text{s}^{-1}$. When experimental data for L_{tot} are available, g_l can be determined.

4.2.3 Nuclei Number Density

There are two important limiting cases of isothermal crystallization, namely the instantaneous nucleation, where the nuclei are there from the beginning, and the sporadic nucleation, where the number of nuclei increases linearly with time.

For quiescent crystallization under a constant temperature, in the case of instantaneous nucleation with a constant number density N_0, one obtains $N_q(t) = N_0 H(t)$ for the activated quiescent nuclei number density, where $H(t)$ is the Heaviside unit step function, zero for $t < 0$ and unity for $t \geq 0$. Then the rate of the nuclei number density is $\dot{N}_q = N_0 \delta(t)$, with $\delta(t)$ being the Dirac delta function concentrated at $t = 0$. Equations 4.1 and 4.3 lead to the familiar Avrami equation:

$$\alpha(t) = 1 - \exp\left(-\frac{4\pi}{3} N_0 G^3 t^3\right). \tag{4.10}$$

For the case of sporadic nucleation, assuming $\dot{N}_q =$ constant, one has

$$\alpha(t) = 1 - \exp\left(-\frac{\pi}{3} \dot{N}_q G^3 t^4\right). \tag{4.11}$$

According to Koscher and Fulchiron (2002), for quiescent instantaneous nucleation, the value of N_0 has an exponential dependence on the degree of supercooling ΔT:

$$\ln N_0 = a_N \Delta T + b_N, \tag{4.12}$$

where a_N and b_N are constants to be determined by fitting the equation to experimental data of $N_0(T)$. In case N_0 cannot be counted by optical observations, it may be evaluated by $N_0 = 3 \ln 2 \big/ (4\pi G^3 t_{1/2}^3)$, where $t_{1/2}$ is the half crystallization time, corresponding to the time when the relative crystallinity $\alpha = 0.5$. The data of half crystallization time under quiescent conditions can be measured from differential scanning calorimetry (DSC) experiments.

The flow-induced nuclei number takes the following form as suggested by Eder and Janeschitz-Kriegl (1997):

$$\dot{N}_f + \frac{1}{\lambda_N} N_f = f, \tag{4.13}$$

where λ_N is a nucleation relaxation time, which, according to Eder and Janeschitz-Kriegl, has a large value and may vary with temperature, and f is a function of flow variables.

Several expressions for the function f have been proposed by different authors based on different assumptions about the driving force for the enhancement of nucleation. The suggested driving forces include shear rate, recoverable strain, the first normal stress difference, the change in free energy induced by flow, the effect of the combination of shear rate and strain, etc.

Eder and Janeschitz-Kriegl (1997) consider a function of shear rate and assume that

$$f = g_n(T) \left(\frac{\dot{\gamma}}{\dot{\gamma}_c}\right)^2, \tag{4.14}$$

where $\dot{\gamma}_c$ is a critical shear rate for the flow-induced nuclei generation, and g_n is a parameter for the nucleation rate.

Eder and Janeschitz-Kriegl's equation is based on the kinematics of the flow and not on the dynamics of molecules experiencing the kinematics. Zuidema (2000) and Zuidema et al. (2001) replaced the shear rate with the second invariant of the deviatoric part of the recoverable strain tensor. Koscher and Fulchiron (2002) considered the first normal stress difference as a driving force for nucleation.

Other authors, for example, Coppola et al. (2001) and Zheng and Kennedy (2004), considered the flow-induced change in free energy as a driving force. The starting point to construct a formulation is based on theories of Lauritzen and Hoffman (1960) and Ziabicki (1996) for the quiescent nucleation rate, which is then extended to include the flow-induced change of free energy. Coppola et al. (2001) propose the following rate function for the total nucleation rate:

$$\dot{N} = C_0 k_B T(\Delta F_q + \Delta F_f)\exp\left(-\frac{E_a}{k_B T}\right)\exp\left(-\frac{K_n}{T(\Delta F_q + \Delta F_f)^n}\right), \tag{4.15}$$

where C_0 includes energetic and geometrical constants, ΔF_q is the Gibbs free energy difference between melt and crystalline phase under quiescent conditions, ΔF_f is the flow-induced free energy change per unit volume (measured in J/m^3), k_B is the Boltzmann constant, T is the absolute temperature, E_a is the activation energy of the supercooled liquid-nucleus interface, and K_n is a constant containing energetic and geometrical factors of crystalline nucleus. The power index n, also appearing as a subscript for K, accounts for the temperature region where the homogeneous nucleation occurs.

Zheng and Kennedy (2004) proposed the following form for function f in Eq. 4.13

$$\begin{aligned} f(\Delta F_f, T) =& C_0 k_B T\exp\left[-\frac{U^*}{R_g(T - T_\infty)}\right] \\ &\times\left\{(\Delta F_q + \Delta F_f)\exp\left(-\frac{K_g}{T\left[(1+\theta\Delta F_f)T_m^0 - T\right]}\right)\right. \\ &\left.-\Delta F_q\exp\left[-\frac{K_g}{T(T_m^0 - T)}\right]\right\}, \end{aligned} \tag{4.16}$$

where θ is given by $\theta = T_m^0/(\Delta H_0 T)$ with ΔH_0 being the latent heat of crystallization (in J/m^3). The FENE-P model has been used to calculate the flow-induced change in free energy and the subsequent nucleation enhancement. For low shear rates, the change in free energy due to flow is (Tanner and Qi 2005a)

$$\Delta F_f \approx \frac{1}{2} n_0 k_B T\left(\frac{b}{b+3}\right)^2 (\lambda\dot{\gamma})^2, \tag{4.17}$$

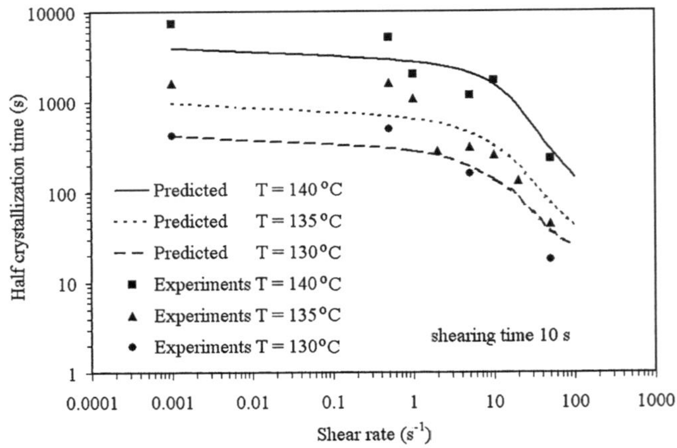

Fig. 4.3 The dependence of the half crystallization time on shear rate and temperature for a sample of iPP. Curves are model predictions. Symbols are experimental results obtained from Linkam shearing hot stage experiments (From Zheng and Kennedy (2004), with permission from American Institute of Physics for the Society of Rheology)

and for high shear rates, one has

$$\Delta F_f \approx \frac{1}{3} n_0 k_B T (b-2) \ln(\lambda \dot{\gamma}), \tag{4.18}$$

where n_0 is the number density of the molecules (the $n_0 k_B T$ is a quantity normally thought as being folded into a shear modulus that can be estimated or measured), and b is the FENE-P model parameter.

Zheng and Kennedy (2004) have shown that the model is capable of describing the Linkam shearing hot stage experiments for the dependence of the half crystallization time on shear rate and temperature reported by Koscher and Fulchiron (2002). Results are shown in Fig. 4.3.

Tanner et al. (Tanner 2003; Dai et al. 2006a; Tanner and Qi 2005, 2009) introduced a simple form for f in shear flows:

$$f = A|\dot{\gamma}|^p \gamma, \tag{4.19}$$

where A and p are constants, $\gamma = \dot{\gamma} t_s$ is the strain and t_s is the shearing time. Wassner and Maier (2000) have measured the viscosity as a function of time under isothermal steady shearing at various shear rates for an isotactic polypropylene. Equation 4.14 fits the experimental data very well when $p = 1.04$. The equation also shows a good agreement with small-amplitude oscillatory experiments. This model may be formulated in an invariant manner by defining $\dot{\gamma} = (2\mathbf{D} : \mathbf{D})^{1/2}$ and $\gamma = (\text{tr}\mathbf{C} - 3)^{1/2}$, respectively (see Eqs. 2.12 and 2.14) and thus can be used in general flows in principle. Boutaous et al. (2010) have applied Eq. 4.19 in a non-isothermal crystallization simulation.

Material characterization under controllable flow conditions are usually carried out in rheometers and shearing hot stage. For example, in experimental studies reported by Hadinata et al. (2005), a parallel plate rotational rheometer, the Advanced Rheometric Expansion System (ARES), was used for viscosity measurements, and a Linkam CSS450 hot stage, as described by Mackley et al. (1999), was used in conjunction with a light intensity measuring device to observe the evolution of morphology.

4.2.4 Molecular Orientation

Mendoza et al. (2003) studied experimentally the influence of processing conditions on the spatial distribution of the molecular orientation in injection molded isotactic polypropylene (iPP) plates. They found that the anisotropy of injection molded semi-crystalline polymers is governed by the orientation of the crystalline phase, and the distribution of the orientation strongly depends on the shear rate. Doufas et al. (2000), followed by Zheng and Kennedy (2001), have applied a rigid dumbbell model to simulate crystalline orientation in injection molded semi-crystalline polymers. The model reads, in the form used by Zheng and Kennedy (2001, 2004):

$$\lambda_c\left(\frac{\Delta\langle\mathbf{pp}\rangle}{\Delta t}+2\mathbf{L}:\langle\mathbf{pppp}\rangle\right)+\langle\mathbf{pp}\rangle=\frac{1}{3}\mathbf{I}, \tag{4.20}$$

where $\mathbf{p}$ is the unit orientation vector, λ_c is the orientation relaxation time for the tumbling motion of $\mathbf{p}$. The angular bracket denotes the ensemble average with respect to the probability density function of the orientation state.

Mendoza et al. (2003) used the Hermans orientation factor describe the measured results. The orientation factor is defined by

$$f_c=\frac{3\langle\cos^2\theta\rangle-1}{2}, \tag{4.21}$$

where θ is the angle between the principal direction of the flow and the molecular chain axis of the crystal. In general, $f_c=-0.5$ corresponds to perfect alignment perpendicular to the flow direction, $f_c=0$ corresponds to random orientation, and $f_c=1$ corresponds to perfect alignment in flow direction. Note that $\langle\cos^2\theta\rangle$ can be expressed by $\langle p_1p_1\rangle$. Thus we can write the orientation factor as

$$f_c=\frac{3\langle p_1p_1\rangle-1}{2}. \tag{4.22}$$

Simulation results of Zheng and Kennedy (2001) compared with experiments of Mendoza et al. (2003) are shown in Fig. 4.4. The overall trend of the orientation distribution is captured. The shearing region exhibits the highest level of orientation, which is solidified during filling. The orientation in the 1 mm-thick plate is

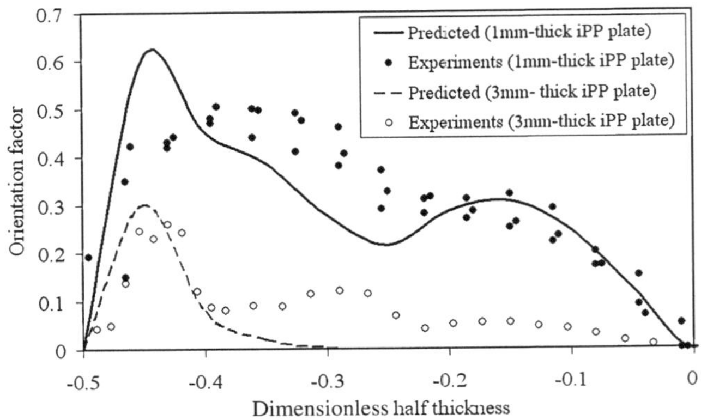

Fig. 4.4 Predicted crystalline orientation factor (Zheng and Kennedy 2001) compared with experimental data (From Mendoza et al. (2003), with permission from Elsevier)

higher than in the 3 mm-thick plate. A gap-wise profile with a second hump is predicted for the 1 mm-thick plate. The second hump corresponds to the orientation frozen in during the post-filling. In the post-filling stage, as the solid/melt interface moves inward, a small flow rate in the narrower channel may still result in a considerable deformation. In addition, due to the decreasing temperature with time, even small shear rates could introduce high stresses and leads to a high level of orientation.

4.3 Effect of Crystallization on Physical Properties

4.3.1 Effect of Crystallization on Rheology

The injection molding process includes both molten polymer and solidification phase, where flow and crystallization may take place simultaneously. Therefore, for modeling injection molding, a rheological relation connecting the relative crystallinity α to flow parameters is needed.

4.3.1.1 Two-Phase Model

Doufas et al. (1999, 2000) proposed a two-phase model based on a modified Giesekus model for the amorphous melt phase and a rigid dumbbell model for the semi-crystalline phases. In the modified Giesekus model the relaxation time is a function of relative crystallinity α:

$$\lambda_a(\alpha, T) = \lambda_a(0, T)(1 - \alpha)^2, \tag{4.23}$$

where the subscript a refers to the amorphous phase. The total extra stress tensor of the system is give by the additive rule:

$$\tau = \tau_a + \tau_{sc}, \tag{4.24}$$

where $\boldsymbol{\tau}_a$ is the extra stress contributed by the amorphous phase calculated from the a modified Giesekus model, while $\boldsymbol{\tau}_{sc}$ is the stress contributed by the semi-crystalline phase calculated from the rigid dumbbell model.

This two-phase model is able to predict the locking-in of stresses and the orientation of crystallites.

Zheng and Kennedy (2004) followed the approach of a two-phase model, but in their model, the effect of crystallinity is not built-into the amorphous parameters. They view the system as a suspension of crystalline phase in a matrix of amorphous phase. The physical properties of the amorphous phase are independent of α, while the viscosity of the whole system is a function of the relative crystallinity.

However, as pointed out by Tanner (2003), the stress law of (4.24) over simplifies the real picture of structure. The additive rule envisages parallel components of amorphous and crystalline phase at each point. Photographs of crystallizing polypropylene such as those shown by Koscher and Fulchiron (2002) and many others do not support this assumption.

4.3.1.2 Viscosity Model

Tanner and Qi (2005) treat viscosity of the whole system as a function of the relative crystallinity through a suspension-like expression:

$$\frac{\eta}{\eta_a} = \left(1 - \frac{\alpha}{A}\right)^{-2}, \quad \alpha < A, \tag{4.25}$$

where η_a is the viscosity of the amorphous melt phase, A is a constant depending on the crystal geometry, taking a value of 0.4–0.68.

Zheng and Kennedy (2004) proposed an alternative equation

$$\frac{\eta}{\eta_a} = 1 + \frac{(\alpha/A)^{\beta_1}}{(1 - \alpha/A)^{\beta}}, \quad \alpha < A, \tag{4.26}$$

where β and β_1 are parameters to be determined by experiments, A is set to 0.44, corresponding to rough and compact inclusions.

Both of above models describe an upturn in viscosity, and show that $\eta/\eta_a \rightarrow \infty$ when $\alpha \rightarrow A$.

Pantani et al. (2001b) suggest the following model

$$\frac{\eta}{\eta_a} = 1 + \beta \exp\left(-\frac{\beta_1}{\alpha^{\beta_2}}\right), \tag{4.27}$$

The following expression

$$\frac{\eta}{\eta_a} = \exp\left(\beta\alpha^{\beta_1}\right) \tag{4.28}$$

has been adopted by Zuidema et al. (2001) and Hieber (2002).

Equations 4.27 and 4.28 also describe the upturn in viscosity. When $\alpha \to 1$, both models show that the viscosity levels off and reaches a constant value.

4.3.2 *Effect of Crystallization on Pressure–Volume–Temperature Relations*

In the governing equations for the injection molding simulation (as discussed in Chap. 3), several important polymer material properties, such as density, compressibility, and thermal expansion coefficient, are related to the pressure–volume–temperature (PVT) relation. Section 2.4 has shown the Tait equation, which includes several parameters to be determined from experimental data. For most widely used laboratory PVT apparatuses, data are measured under quasi-equilibrium state conditions, with few exceptions (e.g., van der Beek et al. 2006). Such conditions, however, never hold in real processing. In injection molding, polymers are subjected to flow and highly non-equilibrium cooling. For semi-crystalline polymers, the flow and thermal history strongly influence crystallization which has a significant effect on the evolution of specific volume. Figure 4.5 shows a PVT diagram where the transition zone shifts to lower temperatures with increasing cooling rate. The influence of flow on specific volume will be opposite, shifting the transition zone up to higher temperatures with increasing pressure. Experimental evidence has been reported by Luyé et al. (2001) and Forstner et al. (2009).

The specific volume v of a semi-crystalline polymer is a combination of the amorphous and crystalline structures. From the mass balance, one obtains

$$\frac{1}{v} = \frac{1 - \alpha\chi_\infty}{v_a} + \frac{\alpha\chi_\infty}{v_c}, \tag{4.29}$$

where v_a is the specific volume of the amorphous phase, v_c is the specific volume of the pure crystalline phase, and χ_∞ is the ultimate absolute crystallinity. The product of the relative crystallinity and the ultimate absolute crystallinity $\alpha\chi_\infty$ denotes the volume fraction of crystallized material. If we assume that, in the solid zone of the PVT data, the crystallization has reached the ultimate state, i.e., the relative crystallinity is equal to one, we have the following expression for the specific volume v_s:

$$\frac{1}{v_s} = \frac{1 - \chi_\infty}{v_a} + \frac{\chi_\infty}{v_c}. \tag{4.30}$$

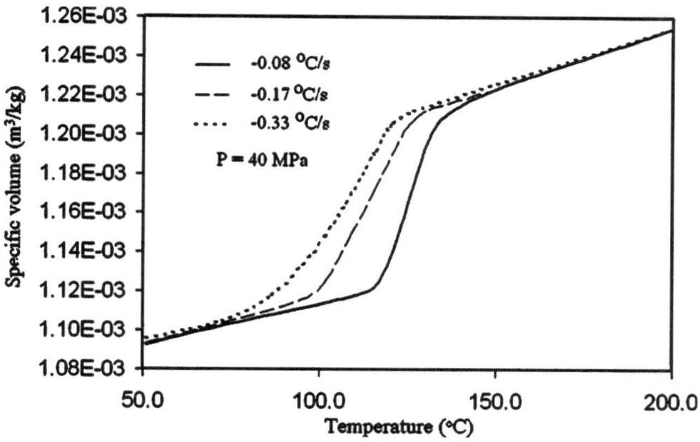

Fig. 4.5 Effect of cooling rate on PVT diagram (Reproduced from Luyé et al. (2001), with permission from John Wiley and Sons)

Eliminating v_c from Eqs. 4.29 and 4.30, we obtain

$$\frac{1}{v} = \frac{1-\alpha}{v_a} + \frac{\alpha}{v_s}. \tag{4.31}$$

Equation 4.31 is useful in simulations. Standard PVT data and the Tait equation can be used for calculation of v_a at melt zone and v_s at solid zone, while the specific volume at transition zone can be calculated using crystallization kinetics and a linear interpolation between $1/v_a$ and $1/v_s$ using Eq. 4.31.

It should be noted that some authors (e.g., Luyé et al. 2001) write the equation as $v = (1-\alpha)v_a + \alpha v_s$, where α is defined as a relative mass crystallinity (rather than the relative volumetric crystallinity as we have used); the expression is equivalent to Eq. 4.31.

4.3.3 Effect of Crystallization on Thermal Conductivity

Thermal conductivity is an important parameter used for material thermal calculations in the injection molding simulation. A variation in the thermal conductivity will alter the cooling rate and hence cause a variation in the temperature; this in turn will cause the viscosity, pressure, and frozen layer to vary. Classical Fourier theory, which assumes thermal conductivity to be a constant scalar value depending only on temperature alone, is inadequate to describe the heat conduction in deformed molten polymer. Van den Brule (1989, 1990) and van den Brule and O'Brien (1990) suggested that heat transport mechanisms along the backbone of a polymer chain are more efficient than those between neighboring chains. Hence, the orientation of polymer chain segments induced by flow leads to an anisotropic thermal conductivity. To describe anisotropic thermal conduction, one has to consider the non-Fourier law (see also Huilgol et al. 1992) with the thermal conductivity being written in a tensor form as

$$q_i = k_{ij} \frac{\partial T}{\partial x_j}. \tag{4.32}$$

Van den Brule proposed a connection between the thermal conductivity tensor and the total stress tensor as follows:

$$k_{ij} - \frac{1}{3} k_{kk} \delta_{ij} = C_t k_0 \left(\sigma_{ij} - \frac{1}{3} \sigma_{kk} \delta_{ij} \right), \tag{4.33}$$

where k_0 is the equilibrium scalar thermal conductivity, δ_{ij} is the unit tensor, and C_t is the stress-thermal coefficient.

Measuring the flow-dependent thermal conductivity is difficult. Nevertheless, some experimental results were reported by Venerus et al. (1999, 2000, 2001 and 2004) and Schieber et al. (2004) for amorphous polymers. The results indeed support the theory of van den Brule. In addition, they found that the product of stress-thermal coefficient C_t and the melt plateau modulus G_N is a nearly universal value, i.e., $C_t G_N \approx 0.03$.

What happens to semi-crystalline polymers is still to be explored. Dai and Tanner (2006b) suggested that the Van den Brule theory might be used for sheared semi-crystalline materials, with an enhanced value of C_t depending on the increased crystallinity. The idea has been applied by Zheng and Kennedy (2006) and Zheng et al. (2010) in the injection molding simulation. They further assume k_0 as a function of the relative crystallinity given by:

$$\frac{1}{k_0(\alpha, T)} = \frac{\alpha}{k_0^{(s)}(T)} + \frac{1 - \alpha}{k_0^{(a)}(T)}, \tag{4.34}$$

where α is the relative crystallinity, $k_0^{(s)}(T)$ is the equilibrium scalar thermal conductivity at solid state, and $k_0^{(a)}(T)$ is the equilibrium scalar thermal conductivity at melt state. Measurement of k_0 at multiple temperatures over the range including both the solid and melt states has been reported by Speight et al. (2010).

4.4 Influence of Colorants

Most studies on crystallization have considered virgin polymers without additives. However, in many injection-molded products, various additives such as colorants are usually added and these are known to affect the crystallization behavior. The topic has been recently tackled by Hadinata et al. (2008), Zheng et al. (2008, 2010), Lee Wo and Tanner (2010), and Zhu et al. (2009). These authors investigated the FIC behavior of an injection-molded isotactic polypropylene (iPP) mixed

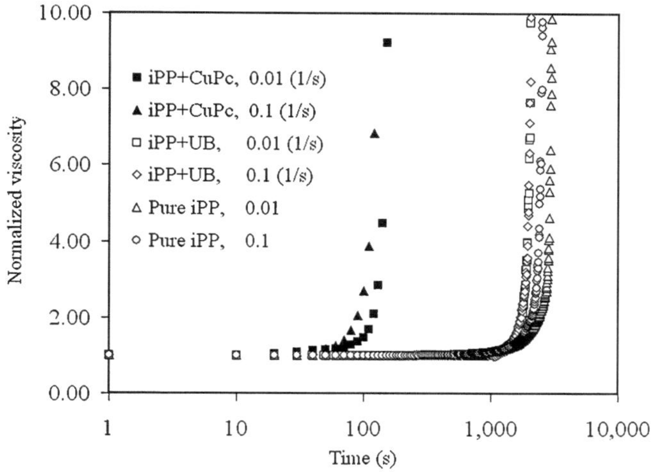

Fig. 4.6 Experimental normalized viscosity against time for the pure iPP and the iPP mixed with CuPc and UB colorants at $T = 140°C$ and $\dot{\gamma} = 0.01\ s^{-1}, 0.1\ s^{-1}$. The concentration of either colorant is 0.8% by weight (provided by Dr. Duane Lee Wo)

with colorant additives. Two types of blue colorants were used in the study: one is the Ultramarine Blue composed of Sodium Alumino Sulpho Silicate (UB) and the other is the PV Fast Blue composed of Cu-Phthalocyanine (CuPc).

Lee Wo and Tanner (2010) have carried out a steady shear experiment with a parallel plate rheometer to monitor the effect of crystallization kinetics on viscosity. Typical experimental results of the normalized viscosity $\eta(t)/\eta_a$ against the time t at a temperature 140°C and two different shear rates, of 0.01 s^{-1} and 0.1 s^{-1}, are shown in Fig. 4.6, where three samples were used: (i) virgin iPP, (ii) iPP filled with 0.8% (by weight) CuPC colorant and (iii) iPP filled with 0.8% (by weight) UB-colorant. The CuPc colorant dramatically reduces the time of the upturn in the viscosity. Since the upturn in the viscosity indicates the onset of crystallization, the results suggest that the CuPC colorant behaves like an excellent nucleating agent. The UB colorant showed a small effect. Zhu et al. (2009) attribute the difference to the effects of the surface patterns, pigment sizes, degree of aggregation and dispersion of the colorants. The CuPC colorant, which has the flat surface full of a nonuniform size distribution, activates effectively the heterogeneous nucleation. In contrast, the UB colorant, which has the curved surface full of a uniform size distribution, has failed to induce the heterogeneous nucleation.

To include the details of colorants in the numerical modeling is difficult, if not impossible. Tactically, in a practical computation, the basic kinetic model for materials with and without colorants can remain the same, while the model parameters should be carefully determined for each colored polymer such that the nucleation and growth trends are correlated with the colorant effects.

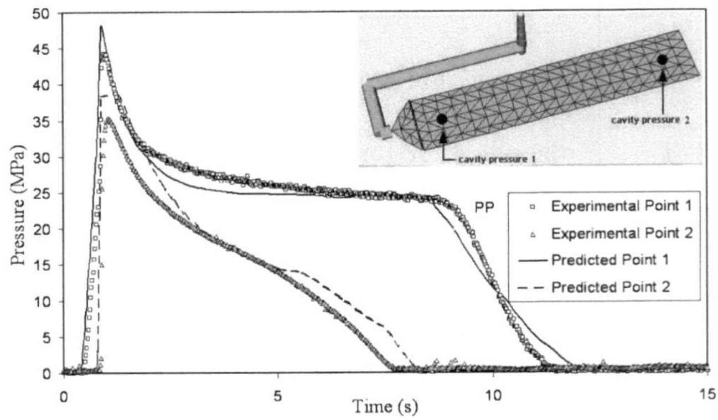

Fig. 4.7 Measured and predicted pressure profiles for the virgin iPP (From Zheng et al. (2010), with permission from Springer Science + Business Media B.V.)

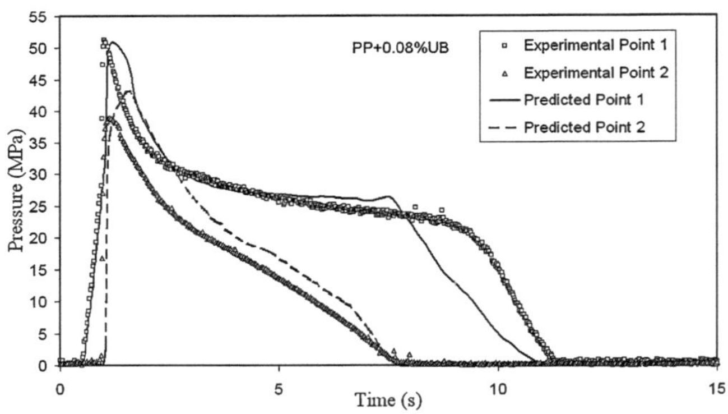

Fig. 4.8 Measured and predicted pressure profiles for the iPP with UB colorant (From Zheng et al. (2010), with permission from Springer Science + Business Media B.V.)

Results of numerical simulation and experiments of injection molding for isotactic polypropylene (iPP) with and without colorants have been reported by Zheng et al. 2008, 2010). Cavity pressure profiles for injection-molded plates of pure iPP, iPP with UB colorant and iPP with CuPC colorant are shown in Figs. 4.7, 4.8 and 4.9, respectively. In spite of the fact that the concentration of the color pigments is only 0.8% by weight, both experimental and predicted results show that the CuPC colorant causes a distinctly earlier decay in pressure. In addition, shrinkage experiments and simulation show a high degree of anisotropic shrinkage exhibited by the CuPc-colored iPP, while the pure iPP and UB-colored iPP do not result in highly anisotropic shrinkage. These results reveal the necessity of incorporating the colorant effect in the simulation of flow-induced crystallization in injection molding.

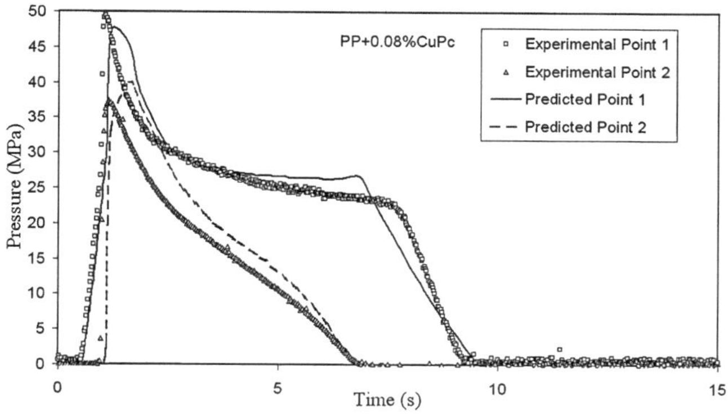

Fig. 4.9 Measured and predicted pressure profiles for the iPP with CuPc colorant (From Zheng et al. (2010), with permission from Springer Science + Business Media B.V.)

4.5 Molecular Dynamics Simulation

In the previous sections we have dealt with some theories to solve problems for polymer crystallization in processing. An alternative possibility is to use molecular-scale simulation to simulate directly the original system by first principles, rather than trying to obtain a simplified closed form expression to describe the behavior of the system. Molecular dynamics (MD) simulation is among the molecular-scale simulation methods that integrates Newton's equations of motion for a set of molecules (Allen and Tildesley 1989; Jabbarzadeh and Tanner 2006). The thermal energy in an MD simulation is just the average kinetic energy of the atoms. In the MD simulation, the equations are deterministic, and the Brownian motion of the molecules is produced by a direct simulation of a huge number of intermolecular collisions, very similar to the case in real fluids. It is this feature that distinguishes MD from other molecular simulation methods such the Brownian dynamics and the Monte Carlo methods. Jabbarzadeh and Tanner (2009, 2010) have used the MD to study the crystallization of polymers at rest, under flow and post-flow conditions.

In the work of Jabbarzadeh and Tanner, linear polyethylene $CH_3(CH_2)_nCH_3$ was modeled using a united atom model; for example, they have recently simulated a model polyethylene $C_{162}H_{326}$ (Jabbarzadeh and Tanner 2010). The groups of CH_2 and CH_3 were treated as single interaction sites. Intra-molecular architecture was used in the model, including bond stretching, angle bending and dihedral (torsional) potentials. Then the molecular positions, velocities and trajectories were calculated from a set of equations as the following:

$$\dot{\mathbf{r}}^{(i)} = \frac{\mathbf{p}^{(i)}}{m^{(i)}} + \mathbf{e}_1 \dot{\gamma} r_3^{(i)} + \dot{\varepsilon}\mathbf{r}^{(i)}, \tag{4.35}$$

$$\dot{\mathbf{p}}^{(i)} = \mathbf{F}^{(i)} - \mathbf{e}_1 p_3^{(i)} \dot{\gamma} - \zeta \mathbf{p}^{(i)} - \dot{\varepsilon}\mathbf{p}^{(i)}, \tag{4.36}$$

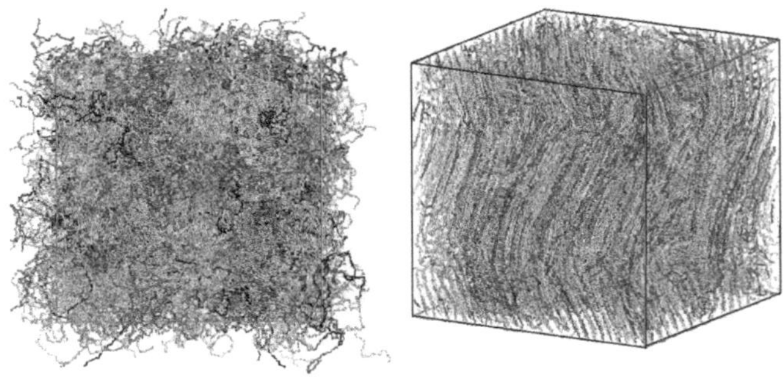

Fig. 4.10 Snapshot of the MD simulation of a model polyethylene, made of 686 molecules, $C_{162}H_{326}$: **a** before crystallization at $T = 393$ K and $P = 1$ atm, and **b** after crystallization, presheared at $\dot{\gamma} = 10^9\ s^{-1}$ for a total strain of $\gamma = 5$ then allowed to crystallize at 350 K for 58 ns (Reproduced from Jabbarzadeh and Tanner (2010), with permission from American Chemical Society)

$$V = 3\dot{\varepsilon}V, \tag{4.37}$$

where $\mathbf{r}^{(i)}$, $\mathbf{p}^{(i)}$, and $m^{(i)}$ are position, momenta and mass of atom i; $\mathbf{e}$ is the unit vector in the flow direction; $\dot{\gamma}$ is the shear rate; $\mathbf{F}^{(i)}$ is the total force applied by all other atoms in the system on atom i; $\dot{\varepsilon}$ is the dilation rate of the simulation box implemented with a Nose–Hoover feedback scheme, and ζ is the Gaussian thermostating multiplier that keeps the temperature constant (Hoover 1985; Evans and Holian 1985); V is the volume of the simulation box. Periodic boundary conditions in all three directions were applied for quiescent simulations. Lees–Edwards sliding brick periodic boundary conditions (Lees and Edwards 1972) were applied for planar shear simulations. Finally, the desired macroscopic properties such as temperature, pressure, density, coefficients of thermal expansion, crystallinity and stresses were calculated from the microscopic information. The simulation is able to monitor the morphological transition from the amorphous structure to the semi-crystalline structure under controlled temperature, pressure, cooling rate and flow conditions for desired polymer system (Fig. 4.10), and has helped to unravel the effect of shearing or pre-shearing on the enhancement of crystallization speed.

Research and development of the MD simulation method is still progressing. Currently, MD simulations are computationally restricted to very tiny time scales (about tens of nanosecond) and very large shear rates ($\geq 10^7\ s^{-1}$). The cost of computer time increases dramatically with increasing number of atoms in the model of molecular chains. The growth of computational power and the development of parallel algorithms will help researchers to simulate crystallization for large systems.

Chapter 5
Flow-Induced Alignment in Short-Fiber Reinforced Polymers

5.1 Concentration Regimes of Fiber Suspensions

A sustained industrial interest has been shown in fiber-filled polymers. When fibers are combined with a polymer matrix that provides cohesion, the fibers become the load bearing component of the composite, and enhance the strength and stiffness of the material. Many articles made from the fiber-reinforced composites are produced by injection molding or compression molding. The thermo-mechanical properties of the end-product highly depend on the fiber orientation distribution induced by the flow of fiber suspension during processing. Therefore, the flow of fiber suspensions needs to be understood in order to predict the fiber orientation distribution and its effects on the end properties of the products.

The concept of a suspension is only meaningful when the typical dimension of the suspended particles is much smaller than the typical dimension of the apparatus. When these two length scales differ by several orders of magnitude, we speak of a suspension rather than a collection of discrete individual particles suspended in a medium.

In a fiber suspension, the fibers are characterized by their aspect ratio and volume fraction. The aspect ratio, a_R, is defined as the length of the particle, l, divided by the diameter of the particle, d. The volume fraction, φ, is defined as the total volume of fibers in a unit volume of suspension. The number density, n, is the number of fibers per unit volume of suspension. As the volume of a single fiber is $\pi d^2 l/4$, we have $\varphi = n\pi d^2 l/4 \approx nd^2 l$. The rheological properties and fiber orientation distribution are the results of the interaction among fibers in the suspension flow. The fiber interactions depend not only on the number of fibers in a certain volume but also on the length of fibers. A group of parameters, such as $nl^3$and ndl^2 are two parameters commonly used to classify the concentration regimes of fiber suspensions. Since $nl^3 = nd^2 l * (l/d)^2 \approx \varphi a_R^2$, and $ndl^2 = nd^2 l * (l/d) \approx \varphi a_R$, the two parameters are equivalent to φa_R^2 and φa_R, respectively.

The concentration of fiber suspensions is classified into three regimes: *dilute*, *semi-concentrated* (or *semi-dilute*) and *concentrated*. A suspension is called dilute

R. Zheng et al., *Injection Molding*, DOI: 10.1007/978-3-642-21263-5_5,

if on average there is less than one fiber in a volume of $V = l^3$. Each fiber can therefore freely rotate without any hindrance from surrounding fibers with three rotational degrees of freedom. This leads to $nl^3 < 1$, or $\varphi a_R^2 < 1$. In the semi-dilute regime, each fiber is confined by $nl^3 > 1$ and $ndl^2 < 1$, equivalent to $1 < \varphi a_R^2 < a_R$, where the average distance between two neighboring fibers is greater than a fiber diameter, but less than a fiber length, and, as a consequence, each fiber effectively has two degrees of rotational freedom. Finally, the suspension with $ndl^2 > 1$ or $\varphi a_R > 1$ is concentrated, where the average distance between fibers is less than a fiber diameter, and therefore fibers cannot rotate independently except around their symmetry axes. Any motion of the fibers must necessarily involve a cooperative effort of all surrounding fibers and fiber–fiber contacts are dominant.

5.2 Evolution Equations

5.2.1 *Jeffery's Orbit*

The starting point of fiber suspension modeling was the work of Jeffery (1922), who modeled the fiber as an inertialess rigid prolate spheroid suspended in a Newtonian fluid. The Jeffery solution for the rate of orientation changes is

$$\dot{p}_i = W_{ij}p_j + \frac{a_r^2 - 1}{a_r^2 + 1}\left(D_{ij}p_j - D_{jk}p_ip_jp_k\right), \tag{5.1}$$

where p_i is a unit vector along the symmetrical axis of the prolate spheroid, $\dot{p}_i$ is the rate of change of the vector p_i and describes the motion following the fiber. If the fibers move with the bulk flow, then $\dot{p}_i \equiv Dp_i/Dt$. D_{ij} is the strain rate tensor and W_{ij} is the vorticity tensor of the flow field, as defined in Eq. 2.8. Note that as $\dot{p}_ip_i = 0$, the magnitude of p_i is preserved in this time evolution. If p_i is initially a unit vector, then it remains a unit vector at all times. There are two physical interpretations for p_i. Firstly, it can be regarded as the local orientation of an individual fiber. Secondly, if there is Brownian motion presents, it represents the averaged configuration. The term $W_{ij}\,p_j$ indicates that p_i rotates with the fluid, while the term $D_{ij}\,p_j$ represents the component of the strain with the fluid. Since p_i has a unit length, the stretching component $D_{jk}\,p_i\,p_j\,p_k$ is subtracted, producing the last term.

For prolate spheroids, a_r is the ratio between the major and the minor axes of the ellipsoid. When applying Eq. 5.1 to cylinders of the length-to-diameter aspect ratio a_R, one may take $a_r \approx a_R$ roughly. Empirical equivalent ellipsoidal axis ratios for cylinders have also been obtained by some authors. For example, Akczurowski and Mason (1968) found that the equivalent ellipsoidal axis ratio a_r for a cylinder of $a_R = 0.86$ is 1.0. They also found that for $a_R > 1.68$, a_r is smaller than a_R, and $l/d < 1.68$, a_r is greater than a_R. Harris and Pittman (1975) obtained from experimental results the following empirical relation: $a_r \approx 1.14a_R^{0.844}$, over the range of $20 < a_R < 400$.

In shear flows of non-interacting fibers, the fibers exhibit a closed periodic rotation known as a Jeffery orbit with a period $T_r = 2\pi\left(a_r + a_r^{-1}\right)/\dot{\gamma}$, where $\dot{\gamma}$ is the generalized strain rate. The Jeffery model is robust for simulations of suspensions of rigid short fibers in the dilute regime, although it has been used to approximate the behavior of suspensions beyond the dilute region (see, for example, Ingber and Mondy 1994). To model realistic suspension flow in injection molding, where the fiber concentrations are usually in the non-dilute regimes, one needs to consider the interactions between fibers.

5.2.2 Orientation Characterization

When a suspension of several fibers needs to be described, the distribution function, also called the probability density function, is the most general way to present the orientation state. This function, $\psi(p, t)$, is defined as the probability to find a fiber orientated in the range between $\mathbf{p}$ and $\mathbf{p} + d\mathbf{p}$ at time t.

The range of $\psi(\mathbf{p}, t)$ is [0, 1]. The integral of this function over all possible values of $\mathbf{p}$ must equal unity, i.e.,

$$\int \psi(p, t)dp = 1. \tag{5.2}$$

Since the directions described by $\mathbf{p}$ and $-\mathbf{p}$ are the same, the distribution function must be even, i.e., $\psi(\mathbf{p}, t) = \psi(-\mathbf{p}, t)$.

The distribution function satisfies the Fokker–Planck equation:

$$\frac{\partial \psi}{\partial t} = \frac{\partial}{\partial p_i}\left[D_r(\delta_{ij} - p_i p_j)\frac{\partial \psi}{\partial p_j} - (\mathscr{L}_{ij}p_j - \mathscr{L}_{jk}p_i p_j p_k)\psi\right], \tag{5.3}$$

where D_r is the diffusion coefficient, and $\mathscr{L}_{ij}$ is an effective velocity gradient, defined as $\mathscr{L}_{ij} = L_{ij} - \xi D_{ij}$, with $\xi = 2/\left(a_r^2 + 1\right)$.

Although the distribution function provides a general description of the orientation state in the suspension, the numerical solution of the Fokker–Planck equation is computationally expensive. One needs a more compact and efficient description of fiber orientation for use in modeling of process. A proper approach is to use orientation tensors (Advani and Tucker 1987). Orientation tensors are defined in term of the ensemble average of the dyadic products of the unit vector $\mathbf{p}$, i.e.,

$$a_{ij} \equiv \langle p_i p_j \rangle = \int p_i p_j \psi d\mathbf{p}, \tag{5.4}$$

$$a_{ijkl} \equiv \langle p_i p_j p_k p_l \rangle = \int p_i p_j p_k p_l \psi d\mathbf{p}, \tag{5.5}$$

where the ensemble average is defined as $\langle * \rangle = \int * \psi(\mathbf{p}, t)d\mathbf{p}$.

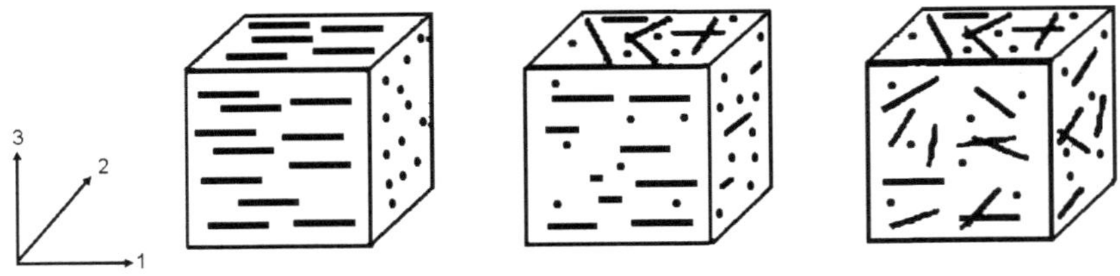

$$a_{ij} = \begin{bmatrix} 1 & 0 & 0 \\ 0 & 0 & 0 \\ 0 & 0 & 0 \end{bmatrix} \quad a_{ij} = \begin{bmatrix} 1/2 & 0 & 0 \\ 0 & 1/2 & 0 \\ 0 & 0 & 0 \end{bmatrix} \quad a_{ij} = \begin{bmatrix} 1/3 & 0 & 0 \\ 0 & 1/3 & 0 \\ 0 & 0 & 1/3 \end{bmatrix}$$

(a) (b) (c)

Fig. 5.1 Some fiber orientation distributions and corresponding second-order orientation tensor components: **a** fully aligned in the 1-direction; **b** random in the 1–2 plane; **c** random in 3-D space

Since ψ is an even function, the average of products of odd number of components of **p** is zero. Also, since the distribution function is normalized (Eq. 5.2) and **p** is a unit vector, one has $a_{kk} = 1$. From the definition, one can also find the symmetry $a_{ij} = a_{ji}$. Therefore, the second-order orientation a_{ij} has only five independent components. Figure 5.1 shows some extreme cases of fiber orientation distributions and corresponding values of the second-order orientation tensor components.

Both the direction and the degree of randomness of the orientation distribution can be indicated by the eigenvectors and the magnitudes of the eigenvalues of a second order tensor. The information can be displayed graphically by an ellipsoid with its axes along the eigenvectors and the magnitudes of its principal axes equal to the eigenvalues. The major axis is then indicating the direction of the preferred fiber orientation, while the roundness of the ellipsoid is a measure of the degree of alignment. In the thin-walled injection molding part, for display purposes, the 3D ellipsoid is projected onto the plane of the part to produce a plane ellipse. This creates a useful representation of orientation distribution. For example, if the orientation is random, then the ellipse is a circle; if the orientation is highly aligned in a particular direction then the ellipse will elongate into a line along the major axis. An example is shown in Fig. 5.2.

5.2.3 Fiber–Fiber Interactions

Folgar and Tucker (1984) have considered the evolution of the orientation state of non-dilute fiber suspensions as a diffusive process. The fiber–fiber interactions are treated as a random force effect superposed with the hydrodynamic effect of the

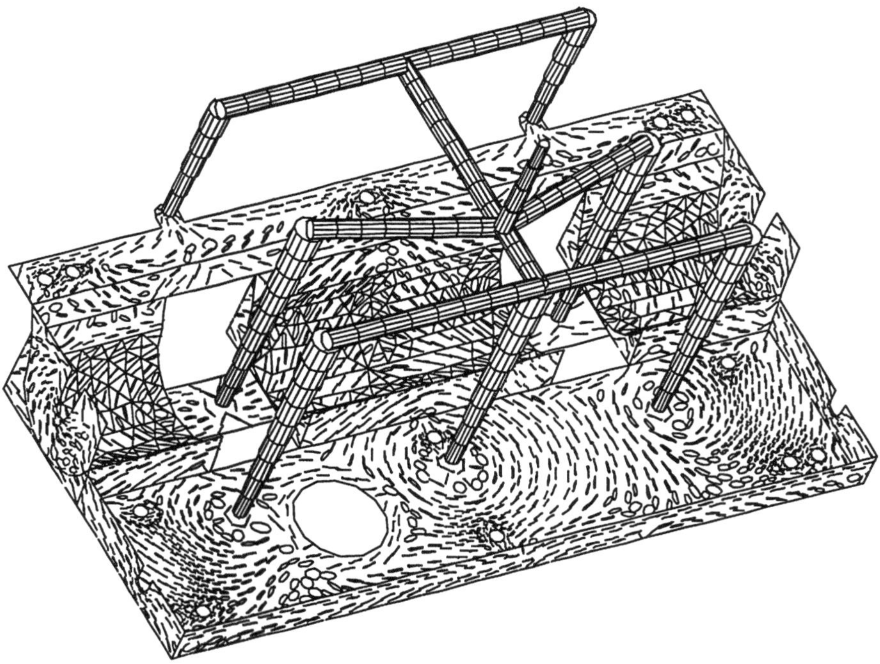

Fig. 5.2 Graphical presentation of predicted fiber orientation distributions (From Zheng et al. (1999) with permission from Elsevier)

fluid flow. This idea is very similar to the incorporation of the Brownian motion in the rigid dumbbell model. Let us rewrite Eq. 5.1 as

$$\dot{p}_i = \mathscr{L}_{ij}p_j - \mathscr{L}_{jk}p_ip_jp_k. \tag{5.6}$$

We then use $\mathbf{F}^{(b)}(t)$ ($F_i^{(b)}(t)$) to denote the resultant interaction force acting on a fiber, and write:

$$\dot{p}_i = \mathscr{L}_{ij}p_j - \mathscr{L}_{jk}p_ip_jp_k + (\delta_{ij} - p_ip_j)F_j^{(b)}(t), \tag{5.7}$$

where factor ($\mathbf{I} - \mathbf{pp}$) in front of $\mathrm{F}^{(b)}(t)$is the statement that only rotational Brownian motion is allowed, since ($\mathbf{I} - \mathbf{pp}$) ensures that the force is always acting perpendicular to $\mathbf{p}$ and keeps it being a unit vector at all times. The statistical model on the fiber interaction is to assume that the fiber interacts with its close neighboring fibers in a random manner. The interaction force $\mathbf{F}^{(b)}(t)$ has zero mean, and the second moments of the interaction forces are given by a delta function:

$$\langle \mathbf{F}^{(b)}(t+s)\mathbf{F}^{(b)}(t)\rangle = 2D_r\delta(s)\mathbf{I}, \tag{5.8}$$

where D_r is the diffusion coefficient, or the strength of the white noise, called the rotary diffusion coefficient, $\delta(s)$is the impulse or Dirac delta function.

Multiplying Eq. 5.6 by tensor $p_i\ p_j$ and then integrating it over the orientation space, one obtains the following evolution equation for the orientation tensor:

$$\frac{\delta a_{ij}}{\delta t} + 2\mathscr{L}_{kl} a_{ijkl} - 2D_r(\delta_{ij} - 3a_{ij}) = 0, \tag{5.9}$$

where the symbol $\delta/\delta t$ represents the Gordon–Schowalter convected derivative, which is equivalent to the upper convected derivative defined with the effective velocity gradient tensor $\mathscr{L}_{ij}$. Equation 5.9 is equivalent to Eq. 2.80 for the rigid dumbbell model, with $\Delta/\Delta t$ and L_{ij} in (2.80) being replaced by $\delta/\delta t$ and $\mathscr{L}_{ij}$, respectively, in (5.9).

Folgar and Tucker assumed

$$D_r = C_I \dot{\gamma}, \tag{5.10}$$

where C_I is called the *interaction coefficient* (or the *Folgar–Tucker constant*). Equation 5.9 with this assumption is called the Folgar–Tucker model, which is widely used for practical injection molding simulations. Problems with the interaction coefficient will be addressed later in Sect. 5.4.

5.3 Closure Approximations

The evolution Eq. 5.9 contains a fourth order tensor a_{ijkl} that is also unknown. This equation is evidently not a closed form. To close the set of evolution equations, a closure approximation has to be introduced to represent the fourth order tensor in term of the second order tensor. There are various closure approximations available.

5.3.1 Linear Closure

The linear closure (Hand 1962) writes

$$\begin{aligned} a_{ijkl} \approx \tilde{a}_{ijkl}^{(Linear)} &= -\frac{1}{35}\left(\delta_{ij}\delta_{kl} + \delta_{ik}\delta_{jl} + \delta_{il}\delta_{jk}\right) \\ &+ \frac{1}{7}\left(a_{ij}\delta_{kl} + a_{ik}\delta_{jl} + a_{il}\delta_{jk} + a_{kl}\delta_{ij} + a_{jl}\delta_{ik} + a_{jk}\delta_{il}\right) \end{aligned} \tag{5.11}$$

For planar orientation the form is the same, but the two coefficients of the terms are $(-1/24)$ and $(1/6)$ respectively. The linear closure is exact for random orientations, but can be dynamically unstable for intermediate to highly oriented states (Advani and Tucker 1990).

5.3.2 Quadratic Closure

The following quadratic closure has been employed by Doi (1981), Lipscomb et al. (1988):

$$a_{ijkl} \approx \tilde{a}_{ijkl}^{(Quad)} = a_{ij}a_{kl}, \tag{5.12}$$

which is exact when the fibers are perfectly aligned, but it can have poor behavior in transient flows or with non-zero interaction coefficients (Advani and Tucker 1990).

5.3.3 Hybrid Closure

The following hybrid closure was proposed by Advani and Tucker (1990):

$$a_{ijkl} \approx \tilde{a}_{ijkl}^{(Hybrid)} = (1-f)\tilde{a}_{ijkl}^{(Linear)} + f\tilde{a}_{ijkl}^{(Quad)}, \tag{5.13}$$

with

$$f = 1 - 27\det(a_{ij}), \text{ for 3-D orientation}, \tag{5.14}$$

and

$$f = 1 - 4\det(a_{ij}), \text{ for planar orientation}. \tag{5.15}$$

The hybrid closure is exact when the fibers are in both random state and perfect alignment. It however tends to accelerate the orientation transients in transient shearing flows.

5.3.4 Composite Closure

The following composite closure was derived by Hinch and Leal (1976):

$$a_{ijkl}D_{kl} \approx a_{ijkl}^{(H-L)}D_{kl} = \frac{1}{5}\left[6a_{ik}D_{kl}a_{lj} - a_{ij}a_{kl}D_{kl} - 2\delta_{ij}a_{km}a_{ml}D_{kl} + 2\delta_{ij}a_{kl}D_{kl}\right]. \tag{5.16}$$

This closure behaves well in some flows but is unrealistic in others.

5.3.5 Orthotropic Fitted Closure

A family of closure approximations called orthotropic closures was developed by Cintra and Tucker (1995). Since the fourth-order orientation tensor a_{ijkl} must take all its directional information from a_{ij}, any objective closure approximation for a_{ijkl} must be orthotropic, having the same principal axes as a_{ij}. Letting $a_{(1)}$, $a_{(2)}$ and $a_{(3)}$ be the principal values of a_{ij}, since $a_{(1)} + a_{(2)} + a_{(3)} = 1$, there are only two independent variables which can be chosen for use in the closure formulation. Cintra and Tucker (1995) further argued that there are only three independent components of a_{ijkl}. They selected a_{1111}, a_{2222} and a_{3333} to construct the basic expressions as follows:

$$\left\{\mathbf{A}^{(Closure)}\right\} = [\mathbf{C}]_{3\times 6}\{\mathbf{a}\}, \tag{5.17}$$

where

$$\left\{\mathbf{A}^{(Closure)}\right\} = \left\{\begin{array}{c} a_{1111}^{(Closure)} \\ a_{2222}^{(Closure)} \\ a_{3333}^{(Closure)} \end{array}\right\}, \tag{5.18}$$

and

$$\{\mathbf{a}\} = \left\{\begin{array}{c} 1 \\ a_{(1)} \\ a_{(1)}^2 \\ a_{(2)} \\ a_{(2)}^2 \\ a_{(1)}a_{(2)} \end{array}\right\}. \tag{5.19}$$

$\left[\bar{\mathbf{C}}\right]_{3\times 6}$ is a 3×6 matrix. Its components were obtained by fitting the formula to numerical solutions of the probability distribution function in a few well-defined flow fields (simple shear, shearing/stretching flows, uniaxial elongation and biaxial elongation). Cintra and Tucker (1995) found that for different ranges of interaction coefficients (C_I), two sets of coefficient matrices are needed. The first one writes:

$$\left[\bar{\mathbf{C}}^{(ORF)}\right]_{3\times 6} = \begin{bmatrix} 0.060964 & 0.371245 & 0.555301 & -0.369160 & 0.318266 & 0.371218 \\ 0.124711 & -0.389402 & 0.258844 & 0.086169 & 0.796080 & 0.544992 \\ 1.228982 & -2.054116 & 0.821548 & -2.260574 & 1.053907 & 1.819756 \end{bmatrix} \tag{5.20}$$

The closure using the above coefficient matrix is labeled as ORF closure, which provides an excellent match to the probability distribution function calculations for $C_I = 0.01$–0.1. However, it exhibits non-physical oscillation for $C_I = 0.001$ in a simple shear flow. Cintra and Tucker (1995) introduced another coefficient matrix with the label ORL for $C_I = 0.001$–0.01, which is given by

$$\left[\bar{\mathbf{C}}^{(ORL)}\right]_{3\times 6} = \begin{bmatrix} 0.104753 & 0.346874 & 0.544970 & -0.563168 & 0.471144 & 0.491202 \\ 0.162210 & -0.451257 & 0.286639 & -0.028702 & 0.864008 & 0.652712 \\ 1.288896 & -2.187810 & 0.899635 & -2.402857 & 1.133547 & 1.975826 \end{bmatrix} \tag{5.21}$$

Further improvements of the orthotropic fitted closure and application to injection molding problems have been reported by VerWeyst (1998) and VerWeyst et al. (1999).

5.3.6 *Natural Closure*

Verleye and Dupret (1993); and Dupret and Verleye (1999) developed a different closure scheme known as the *natural closure*. The closure begins with a general form for a_{ijkl} in terms of a_{ij}:

$$\begin{aligned} a_{ijkl}^{(\mathrm{NAT})} =& \beta_1 S(\delta_{ij}\delta_{kl}) + \beta_2 S(\delta_{ij}a_{kl}) + \beta_3 S(a_{ij}a_{kl}) \\ &+ \beta_4 S(\delta_{ij}a_{km}a_{ml}) + \beta_5 S(a_{ij}a_{km}a_{ml}) + \beta_6 S(a_{im}a_{mj}a_{kn}a_{nl}). \end{aligned} \tag{5.22}$$

where the operator S indicates the symmetric part of its argument,

$$\begin{aligned} S(T_{ijkl}) =& \frac{1}{24}(T_{ijkl} + T_{jikl} + T_{ijlk} + T_{jilk} + T_{klij} + T_{lkij} + T_{klji} + T_{lkji} \\ &+ T_{ikjl} + T_{kijl} + T_{iklj} + T_{kiljl} + T_{jlik} + T_{ljik} + T_{jlki} + T_{ljki} \\ &+ T_{iljk} + T_{lijkl} + T_{ilkj} + T_{likj} + T_{jkil} + T_{kjil} + T_{jkli} + T_{kjli}). \end{aligned} \tag{5.23}$$

The coefficients β_1 to β_6 are polynomial expansions of the second and third invariants of a_{ij}, and chosen to satisfy $a_{ijkk} = a_{ij}$. The polynomial coefficients can be found by matching particular solutions for the exact solutions of the probability distribution function. The idea is similar to the orthotropic fitted closure. However, the natural closure uses analytical solutions of the probability distribution function; the analytical solution is only available for $C_I = 0$ and $a_r \to \infty$.

5.3.7 *Invariant-Based Optimal Fitting (IBOF) Closure*

The IBOF closure proposed by Chung and Kwon (2002) uses the same idea of the natural closure, approximating the fourth-order orientation tensor in terms of the second-order orientation tensor and its invariants. They showed that there are only three independent $\beta_n' s$ among the six $\beta_n' s$ that appear in Eq. 5.22. They chose β_3, β_4 and β_6 as the independent parameters and the rest can be expressed in terms of these three and invariants. Unknown parameters in the polynomial expansions are determined using the method introduced by a least-square optimization fitting technique of various flow data generated from solutions of the probability distribution function for arbitrary C_I and a_r, similar to the approach introduced by Cintra and Tucker (1995) for the orthotropic fitted closure. The IBOF closure thus can be regarded as a hybrid of the natural closure and the orthotropic fitted closure.

5.4 Interaction Coefficient

Now, a remaining problem is how to determine the value of the interaction coefficient. An approach was presented by Bay (1991) based on experimental values of a_{11} in steady simple shear flows for different concentrations. These experimental results were fitted to the numerically calculated a_{11} and the value of C_I that best fits the experimental results was obtained. The polymer matrices used in the experiments were nylon, polycarbonate and polybutylene terephthalate. The data of C_I were then plotted against $a_r\varphi$ to give an empirical relationship

$$C_I = 0.0184\exp(-0.7184\varphi a_r). \tag{5.24}$$

In Bay's experiments, the measured a_{11} value increases as the fiber volume fraction increases, and therefore the empirical equation shows C_I decreasing with increasing φa_r. This is opposed to the trend observed by Folgar and Tucker (1984) for nylon fibers in silicon oil. However, Folgar and Tucker's data were measured in the semi-concentration regime, while Bay's data were measured in the concentrated regime. Bay (1991) and Tucker and Advani (1994) conjectured that there could be a changeover in the fiber–fiber interaction from "disturbances" to "caging" at a certain concentration (around $\varphi a_r = 1$).

It also needs to be mentioned that, in Bay's work, the hybrid closure was used to model the data. With this closure, the values of C_I were found around 0.01 to match the experimental a_{11} values. However, if one uses the orthotropic closure, the data are matched with $C_I = O(0.001)$. Therefore, it is important to notice that, although the measured values of a_{11} are irrelevant to any closure schemes, the data of C_I, and hence the coefficients of the empirical equation (5.25), depend on the closure-scheme chosen.

Ranganathan and Advani (1991) assumed a relationship between C_I and the average inter-fiber spacing a_c for semi-dilute suspensions:

$$C_I = Kl/a_c, \tag{5.25}$$

where l is the fiber length and K is a universal constant of proportionality.

Phan-Thien et al. (2002) proposed the following empirical expression based on calculations of the diffusion coefficient using a direct simulation of fiber suspension dynamics:

$$C_I = 0.03[1 - \exp(-0.224\varphi a_r)]. \tag{5.26}$$

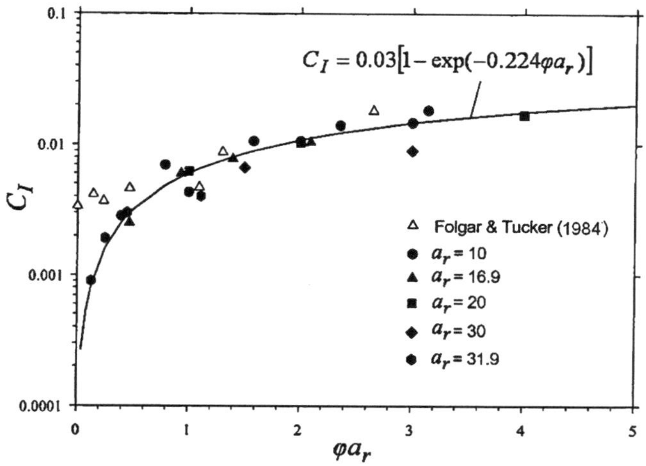

Fig. 5.3 A comparison of the simulated C_I for different aspect ratios (*filled symbols*) with experimental data of Folgar and Tucker (*open symbols*). The *solid line* represents Eq. 5.26 (From Phan-Thien et al. (2002), with permission from Elsevier)

The results of the direct simulation and the empirical equation 5.26 are comparable with the data of Folgar and Tucker (1984), as shown in Fig. 5.3.

There is no compelling reason to assume isotropy in the rotary diffusivity, and in the general case, one may expect an anisotropic interaction coefficient and write it in a tensor form (see, Fan et al. 1998; Phan-Thien et al. 2002). The scale value of C_I shown in Fig. 5.3 can be defined as one-third of the trace of the interaction coefficient tensor. The anisotropic rotary diffusion model will be described in the next Section.

5.5 Modifications to Folgar–Tucker Model

5.5.1 Anisotropic Rotary Diffusion Model

Fan et al. (1998) considered that the diffusion constant could be anisotropic, and they still assumed it proportional to the generalized shear rate, with the constant of proportionality given by a second-order symmetric tensor $\tilde{C}$.

$$\mathbf{D}_r = \tilde{\mathbf{C}}\dot{\gamma}. \tag{5.27}$$

By defining

$$\mathcal{L}_{ij} = \mathscr{L}_{ij} - 3\tilde{C}_{ij}\dot{\gamma}, \tag{5.28}$$

the new orientation evolution equation is written as

$$\frac{\mathcal{D}a_{ij}}{\mathcal{D}t} + 2\mathcal{L}_{kl}a_{ijkl} - 2\dot{\gamma}(\tilde{C}_{ij} - a_{ij}\tilde{C}_{kk}) = 0, \tag{5.29}$$

where the symbol $\mathcal{D}/\mathcal{D}t$ represents the upper convected derivative defined with the velocity gradient tensor L_{ij} being replaced by $\mathcal{L}_{ij}$, that is,

$$\frac{\mathcal{D}a_{ij}}{\mathcal{D}t} = \frac{Da_{ij}}{Dt} - \mathcal{L}_{ik}a_{kj} - \mathcal{L}_{jk}a_{ki}. \tag{5.30}$$

If the diffusivity is isotropic, $\tilde{\mathbf{C}}$ is reduced to an isotropic tensor:

$$\tilde{C}_{ij} = C_I\delta_{ij}, \tag{5.31}$$

where C_I is the Folgar–Tucker interaction coefficient, and Eq. 5.29 is reduced to the standard Folgar–Tucker equation. In the general case, Fan et al. (1998) assume that $\tilde{C}$is a function of the strain rate tensor and its second invariant, given by

$$\tilde{C}_{ij} = c_0\delta_{ij} + c_1\frac{D_{ij}}{\dot{\gamma}} + c_2\frac{D_{ik}D_{kj}}{\dot{\gamma}^2}, \tag{5.32}$$

where the coefficients c_0, c_1 and c_2 can be determined by fitting the model to the numerical data produced by the direct multi-particle simulation. These coefficients can be functions of the fiber aspect ratio and concentration.

Phelps and Tucker (2009) further assumed that $\tilde{\mathbf{C}}$ could also be a function of the orientation tensor, and they proposed the following expression:

$$\tilde{C}_{ij} = c_0\delta_{ij} + c_1a_{ij} + c_2a_{ik}a_{kj} + c_3\frac{D_{ij}}{\dot{\gamma}} + c_4\frac{D_{ik}D_{kj}}{\dot{\gamma}^2}. \tag{5.33}$$

5.5.2 *Reduced-Strain Closure Model*

As experimental evidence has shown that the standard Folgar–Tucker model predicts a faster transient orientation evolution than that observed experimentally, Tucker et al. (2007), Wang et al. (2008) and Phelps and Tucker (2009) have proposed a new evolution equation, i.e., the so-called reduced-strain closure (RSC) model, to slow down the fiber orientation kinetics. Their approach is based on the *spectral decomposition theorem.* The theorem states that if $\mathbf{T}$ is a symmetric second-order tensor, then there is a basis $\{\mathbf{e}_i,\ i = 1, 2, 3\}$ consisting entirely of eigenvectors of $\mathbf{T}$ and the corresponding eigenvalues $\{\lambda_i,\ i = 1, 2, 3\}$ forming the entire spectrum of $\mathbf{T}$, thus $\mathbf{T}$ can be represented by $\mathbf{T} = \sum_{i=1}^{3}\lambda_i\mathbf{e}_i\mathbf{e}_i$.

The theorem allows decomposition of the evolution equation for the second-order tensor into two rate equations for the eigenvalues and eigenvectors of a_{ij}, respectively. Then one modifies the equation for the eigenvalues, and keeps the equation for the eigenvectors unchanged. Finally, a new equation is formed by reassembling the equations. This approach preserves objectivity of the equation. For details about objectivity, see for example Tanner (2000). The resulting RSC equation is

$$\frac{\delta a_{ij}}{\delta t} + 2\mathscr{L}_{kl}\mathrm{A}_{ijkl} - 2\kappa C_I\dot{\gamma}(\delta_{ij} - 3a_{ij}) = 0, \tag{5.34}$$

where

$$\mathrm{A}_{ijkl} = a_{ijkl} + (1-\kappa)(\Lambda_{ijkl} - \mathrm{M}_{ijmn}a_{mnkl}). \tag{5.35}$$

Here the scalar factor κ is a phenomenological parameter that is introduced to reduce the rate of fiber alignment, and its value is to be determined by fitting the fiber orientation or rheological predictions to experimental data. If $\kappa = 1$, the standard Folgar–Tucker equation is recovered. The fourth-order tensors Λ_{ijkl} and M_{ijkl} can be calculated from eigenvalues and eigenvectors of the second-order orientation tensor a_{ij}. Let $\mathbf{e}_i$ and λ_i $(i = 1, 2, 3)$ be the eigenvectors and the corresponding eigenvalues of a_{ij}, respectively, the fourth-order tensors $\mathbf{\Lambda}$ and $\mathbf{M}$ are defined as

$$\mathbf{\Lambda} = \sum_{i=1}^{3} \lambda_i \mathbf{e}_i\mathbf{e}_i\mathbf{e}_i\mathbf{e}_i, \tag{5.36}$$

and

$$\mathbf{M} = \sum_{i=1}^{3} \mathbf{e}_i\mathbf{e}_i\mathbf{e}_i\mathbf{e}_i. \tag{5.37}$$

A similar model has also been proposed by Férec et al. (2009) independently.

5.6 Rheological Equations for Fiber Suspensions

The central problem in the rheology of suspensions is to develop suitable constitutive relations to relate the fiber orientation to the stress tensor. In general, for a suspension of fibers in a Newtonian fluid, the constitutive equations have two parts contributing to the extra stress of the suspension:

$$\tau_{ij} = \tau_{ij}^{(s)} + \tau_{ij}^{(p)}, \tag{5.38}$$

where $\tau_{ij}^{(s)}$ is the viscous contribution of the suspending fluid and $\tau_{ij}^{(p)}$ is the particle-contributed stress. The difference among various constitutive models is the expression of the particle-contributed stresses.

5.6.1 Transversely Isotropic Fluid (TIF) Model

Ericksen (1960) proposed a model for the transversely isotropic fluid (TIF), given by:

$$\tau_{ij}^{(p)} = 2\eta_s\varphi\left[A_1 D_{kl}a_{ijkl} + A_2\left(D_{ik}a_{kj} + a_{ik}D_{kj}\right) + A_3 D_{ij} + D_r A_4 a_{ij}\right], \tag{5.39}$$

where η_s is the solvent viscosity, φ is the particle volume fraction, D_r is the rotational diffusivity, and A_1 to A_4 are material constants depending on the aspect ratio a_r. For $a_r \gg 1$ (rod-like), the asymptotic values of these constants are (Lipscomb et al. 1988):

$$A_1 = \frac{a_r^2}{2(\ln 2a_r - 1.5)}, \tag{5.40}$$

$$A_2 = \frac{6\ln 2a_r - 11}{a_r^2}, \tag{5.41}$$

$$A_3 = 2, \tag{5.42}$$

$$A_4 = \frac{3a_r^2}{\ln 2a_r - 0.5}. \tag{5.43}$$

The TIF model is only appropriate for dilute suspensions.

5.6.2 Dinh–Armstrong Model

The Dinh–Armstrong model (Dinh and Armstrong 1984) assumes that the fiber aspect ratio is infinite, and the particles-contributed stress in a homogeneous flow field is given by

$$\tau_{ij}^{(p)} = \varphi\eta_s \frac{\pi l^3 n^3}{6\ln(2H_f/d)} L_{kl}a_{ijkl}, \tag{5.44}$$

where L_{kl} is the velocity gradient tensor, n is the number of particles per unit volume, l and d are the fiber length and diameter, respectively and H_f is the average distance from a fiber to its neighbor, given by $H_f = (nl^2)^{-1}$ for random orientation, and $H_f = (nl)^{-1/2}$ for fully aligned orientation. This model can be applied to semi-concentrated suspensions.

5.6.3 Phan-Thien–Graham Model

The Phan-Thien–Graham model (1991) is a modification of the TIF model. First, the model includes only the a_{ijkl} term, which is the dominant term in the TIP model at large aspect ratios, since $A_1 = O(a_r)$, while $A_2 = O\left(a_r^{-1}\right)$; the Brownian motion is also negligible at large Peclet number [$\mathrm{Pe} = O(\eta_s\dot{\gamma}l^3/k_B T)$]. Secondly, the Phan-Thien–Graham model uses a functional dependence on the volume fraction to replace the linear dependence of the volume fraction in the TIF model, so that it is able to model the transition behavior of concentrated suspensions. In this model, the particles-contributed stress is given by

$$\tau_{ij}^{(p)} = 2\eta_s f(\varphi, a_r) D_{kl} a_{ijkl}, \tag{5.45}$$

where f is a function of the volume fraction and the aspect ratio given by

$$f(\varphi, a_r) = \frac{\varphi a_r^2 (2 - \varphi/\varphi_m)}{4(\ln 2a_r - 1.5)(1 - \varphi/\varphi_m)^2}, \tag{5.46}$$

in which φ_m denotes the maximum volume packing, which can be evaluated by

$$\varphi_m = 0.53 - 0.013a_r, \quad 5 < a_r < 30. \tag{5.47}$$

5.7 Tucker's Flow Classification for Fiber Suspension in Thin Cavities

Lipscomb et al. (1988) have shown that the flow kinematics in complex flows can be strongly affected due to fibers. However, in the injection molding simulation for flows in narrow channels, the effects can be neglected, and an orientation-flow decoupled simulation approach is used. To understand in which situations the decoupled simulation is allowed, we need to classify the flow regimes. Tucker (1991) has identified four regimes of suspension flows. His classification was based on two dimensionless numbers. The first is the *particle number* defined by

$$N_p = \frac{\varphi a_r^2}{3(1 + 2\varphi) \ln Z}, \tag{5.48}$$

with

$$Z = \begin{cases} (\pi/\varphi)^{1/2} & \text{for aligned fibers} \\ \pi/(2\varphi a_r) & \text{for random fibers} \end{cases}. \tag{5.49}$$

The second is a small parameter describing the order of magnitude of out-of-plane orientation given by

$$\delta = \max(\varepsilon, C_I^{1/3}), \tag{5.50}$$

where ε is the slenderness of the gap of the flow channel defined as $\varepsilon = h/L$ (h is the thickness and L is the typical in-plane dimension), and C_I is the interaction coefficient. Equation 5.49 indicates that $\delta \sim \varepsilon$ if the interaction effect is small, while for large interaction coefficients, say, $C_I^{1/3} \gg \varepsilon$, the interaction effect will dominate the out-of-plane orientation. Equation 5.50 suggests that the order of magnitude of δ could never be smaller than $O(\varepsilon)$. This general condition, however, has some exceptions. For example, in a plug flow, the fibers might lie very flat in-plane initially and never be rotated out of plane by the flow. In this special case, it is possible to have $\delta \ll \varepsilon$.

The key dimensionless group to classify the flow regime is $N_p\delta^2$. The four regimes defined by Tucker (1991) are:

Regime I. Decoupled lubrication flow: $N_p\delta^2 \ll 1$, and $\delta \geq \varepsilon$, in which the particle-contributed gap-wise stresses are negligible and hence one is allowed to decouple the flow calculation from the orientation calculation. The lubrication approximation is valid in these regimes.

Regime II. Coupled lubrication flow: $N_p\delta^2 \geq 1$, and $\varepsilon \ll \delta \ll 1$, in which the flow and orientation are coupled, but the lubrication approximation (Hele-Shaw equation) is still applicable. For examples of investigations in this regime, see Chung and Kown (1995, 1996), who incorporated the Dinh–Armstrong model into a coupled analysis of mold filling flow and fiber orientation.

Regime III. General narrow gap flow: $N_p\delta^2 \geq 1$, and $\delta \sim \varepsilon$, where the flow and orientation are coupled, and the lubrication approximation is no longer appropriate, but some simplifications are still possible.

Regime IV. Plug flow with shear boundary layers: $N_p\varepsilon^2 \gg 1$, and $\delta \ll \varepsilon$, where the gap-wise velocity profile is flat with boundary layers of thickness in the order of $N_p^{-1/2}l$.

Consider a set of typical values in the injection molded composites: $\varphi = 10\%$, $a_r = 20$, $C_I = 0.001$and $\varepsilon = 0.01$. From Eqs. 5.48 and 5.50, assuming the fibers are aligned, we obtain $N_p = 6.4$ and $\delta = 0.1$. Then we have $N_p\delta^2 = 0.064$. The flow belongs to regime one, and hence decoupled flow and fiber orientation calculations will be allowed. This will be the case for many, but not all, injection molding simulations for fiber-filled composites.

5.8 Fiber Migration in Inhomogeneous Flow Fields

In injection molding simulation, it is usually assumed that the volume fraction of fiber suspensions is constant in flow fields at all times. However, following the investigations on the particle migration in flow fields (see for example, Gadala-Maria and Acrivos 1980; Leighton and Acrivos 1987a, b; Koh et al. 1994; Mondy et al. 1994), one may expect that fibers can also migrate from the region of high shear rate to that of low shear rate.

To simulate the fiber migration numerically, one can use the diffusive flux model proposed by Phillips et al. (1992), which is given by

$$\frac{\partial \varphi}{\partial t} + u_i \frac{\partial \varphi}{\partial x_i} = \frac{\partial N_i}{\partial x_i}, \tag{5.51}$$

where φ is the volume fraction of the solid phase, u_i is the velocity vector, and N_i is the particle flux, expressed as

$$N_i = k_c a^2 \varphi \frac{\partial}{\partial x_i}(\dot{\gamma}\varphi) + k_\eta a^2 \dot{\gamma} \varphi^2 \frac{\partial \ln \eta_r}{\partial x_i}, \tag{5.52}$$

where k_c and k_η are two empirical parameters determined from experimental data, a is the dimensionless equivalent radius of particles, $\dot{\gamma}$ is the generalized strain rate, and η_r is the relative viscosity define by the ratio of the suspension viscosity to the solvent viscosity ($\eta_r = \eta/\eta_s$). The diffusion equation is subject to a no-flux boundary condition, $N_i n_i = 0$, at the solid boundaries, where n_i is the outward normal unit vector. The initial condition is a uniform volume fraction over the whole domain of suspension fluid.

Fan et al. (2000a) applied this model to simulate the fiber migration in Couette flow and planar Poiseuille flow. The simulation revealed that it takes a long time to achieve a steady state concentration distribution. Since the filling stage of injection molding usually takes very short time to complete, one can expect that the fiber migration effect is not important in injection molding process, although some fiber migrations have been observed in the gate region (Singh and Kamal 1989).

5.9 Brownian Dynamics Simulation

Instead of solving the evolution equation in terms of the orientation tensor, one can simulate the stochastic equation such as Eq. 5.7 for the orientation vector **p** without the need of closure approximations, using the numerical technique for the simulation of stochastic processes (Öttinger 1996) known as the Brownian dynamics simulation. Once trajectories for all fibers are obtained, the orientation tensor can be calculated in terms of the ensemble average of the discrete form:

$$a_{ij} = \frac{1}{N}\sum_{n=1}^{N} p_i^{(n)} p_j^{(n)}, \quad a_{ijkl} = \frac{1}{N}\sum_{n=1}^{N} p_i^{(n)} p_j^{(n)} p_k^{(n)} p_l^{(n)}, \tag{5.53}$$

where the superscript (n) denotes the quantities of the nth fiber and N is the number of fibers. To describe the method, we re-write Eq. (5.7) here:

$$\dot{p}_i = \mathscr{L}_{ij} p_j - \mathscr{L}_{jk} p_i p_j p_k + (\delta_{ij} - p_i p_j) F_j^{(b)}(t). \tag{5.54}$$

We can introduce a vector q_i that satisfies

$$q_i = Q p_i, \tag{5.55}$$

where $Q = (q_k q_k)^{1/2}$. Substituting Eq. 5.55 into 5.54 leads to

$$\dot{q}_i = \mathscr{L}_{ij} q_j + Q F_i^{(b)}(t). \tag{5.56}$$

Different from the molecular dynamics (MD) simulation method (Sect. 4.5), the Brownian dynamics approach does not directly simulate the inter-particle collision. Instead, in the Brownian dynamics, the pseudorandom motion characteristic of the effect of particle–particle interactions is mimicked by a stochastic force generated from random numbers. This makes the Brownian dynamics more efficient than the MD. Here we express the term $F_i^{(b)}$ by a white noise. For a general case where the interactions between fibers are allowed to be anisotropic (Phan-Thien et al. 2000), we have

$$F_i^{(b)} = (2\dot{\gamma})^{1/2}\tilde{C}_{ij}^{1/2}Q\frac{dw_j}{dt}. \tag{5.57}$$

where w_j is the Wiener process, and $\tilde{C}_{ij}$ is the anisotropic interaction coefficient discussed previously. For the case of isotropic rotary diffusion, $\tilde{C}_{ij} = C_I\delta_{ij}$, Eq. 5.56 reduces to

$$F_i^{(b)} = (2\dot{\gamma}C_I)^{1/2}Q\frac{dw_i}{dt}. \tag{5.58}$$

Equation 5.54 can be discretized in time difference:

$$q_i(t+\Delta t) = q_i(t) + \mathscr{L}_{ij}q_j(t)\Delta t + \left(2\dot{\gamma}\tilde{C}_{ij}\right)^{1/2}Q(t)\Delta w_j(t) \tag{5.59}$$

where $\Delta w_j$is the increment of the Wiener process, which is a Gaussian random function. Equation 5.59 can be solved numerically for each particle in terms of variable q_i, and the orientation vector p_i and the orientation tensor a_{ij} are calculated using Eqs. 5.55 and 5.53, respectively. The Brownian dynamics simulation makes it possible to do the multi-scale modeling, that is, the macroscopic flow field is simulated directly from microscopic models without using any closed form of constitutive equations. The stresses are determined from the configurations of the microstructure.

Öttinger (1996) combined the Brownian dynamics simulation technique with finite elements to solve polymer flow problems. The approach has been introduced as the CONNFFESSIT (Calculation of Non-Newtonian Flow: Finite Element and Stochastic Simulation Techniques) approach. Hulsen et al. (1997) extended the approach to the so-called Brownian configuration field (BCF) method, which treats the stochastic equation as the stochastic field equation, and hence avoids the difficulties associated with individual molecule tracking. If the BCF is applied to fiber suspension flows, the vectors **p** and **q** will be functions of space and time. The discrete equation for the time evolution is

$$\begin{aligned} q_i(x_k, t+\Delta t) &+ u_j\frac{\partial q_i(x_k, t+\Delta t)}{\partial x_j}\Delta t \\ &= q_i(x_k, t) + \mathscr{L}_{ij}q_j(x_k, t)\Delta t + \left(2\dot{\gamma}\tilde{C}_{ij}\right)^{1/2}Q(x_k, t)\Delta w_j(t) \end{aligned} \tag{5.60}$$

Fan et al. (1999, 2000a) and Fan (2006) coupled the Brownian configuration field method with the DAVSS (discrete adaptive viscoelastic stress

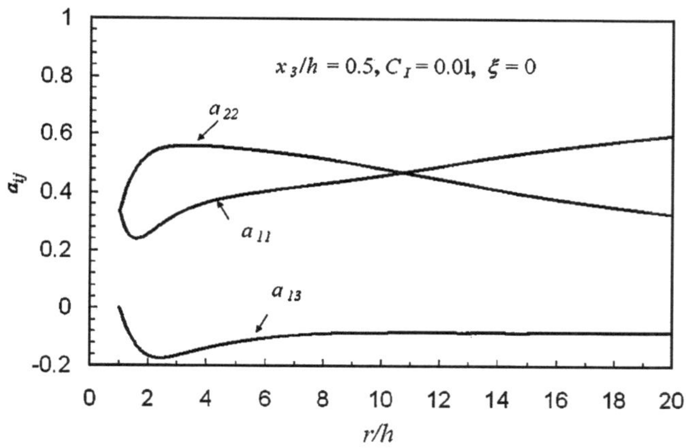

Fig. 5.4 Brownian dynamics simulation results of orientation tensor components distributed along the radial distance of a center-gated disk (From Zheng et al. (2000), with permission from Society of Plastics Engineers Inc.)

splitting) method of Sun et al. (1996, 1998) to solve the fiber motion in complex flows.

Zheng et al. (2000) have demonstrated an example of using the Brownian dynamics simulation to solve injection molding problems. They considered the filling of a center-gated disk with fibers suspending in an isothermal Newtonian fluid. This is a useful benchmark used by many authors (see, for example, Cintra and Tucker (1995), among the others) to test their models or numerical methods. The geometry of the center-gated disk is simple, yet it is often encountered in industry. The flow field is a combination of shear flow and extensional flow. If the fluid fills the cavity at a constant flow rate, the flow decelerates with increasing radial distance, and the in-plane extensional velocity gradient tends to align the fiber in the circumferential direction in the core region of the cavity. Across the thickness, however, the shearing effect competes with the extensional effect and tends to align the fibers with the flow in the radial direction. Moreover, the extension effect becomes weaker at larger radial distances. The total thickness of the disk is 2 h. The predicted orientation tensor components a_{11}, a_{22} and a_{13} along the dimensionless radial distance r/h at the gap-wise location $x_3/h = 0.5$ are plotted in Fig. 5.4. The patterns of orientation tensor component a_{11} across the thickness at the radial distance $r/h = 5$, 10, and 20 are plotted in Fig. 5.5.

5.10 Non-Newtonian Matrix Suspensions

Most suspension theories assume that the suspending medium is a Newtonian fluid. However, in the injection molding process, fibers are dispersed in polymeric liquids, which in most cases are non-Newtonian. Joseph and Liu (1993) have

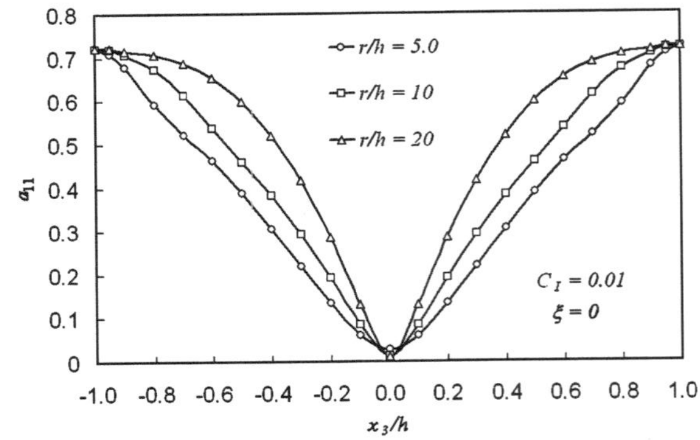

Fig. 5.5 Brownian dynamics simulation results of non-dilute fiber orientation tensor components across the thickness of a center-gated disk, at three different radial positions: $r/h = 5.0$, 10 and 20 (From Zheng et al. (2000), with permission from Society of Plastics Engineers Inc.)

shown that a viscoelastic solvent can dramatically change the nature of a fiber orbit by examining the drop of a needle in Newtonian and viscoelastic fluids.

Phan-Thien and Fan (2002) have used an indirect boundary element method to simulate a single fiber motion in weakly viscoelastic shear flow. This is a first step to understand how much viscoelasticity changes Jeffery's orbit. Based on the numerical information, a modified effective velocity gradient is then introduced into Jeffery's equation to take into account the effect of weak elasticity in flow. Finally, the modified equation is extended further to include the interactions between fibers by adding a random force into the equation, following the approach originally proposed by Folgar and Tucker (1984). This model has been solved by the Brownian dynamics simulation method, and tested against the indirect boundary element result for $C_I = 0$. The new results are: (i) The fiber is spiraling to the vorticity axis in shear flow, driven by the viscoelastic force; (ii) The contribution of viscoelasticity tends to lengthen the period of the spiral orbit; (iii) The fiber-contributed viscosity and first normal stress difference are reduced compared with those for fiber suspensions in the Newtonian fluid. The second normal stress difference is found to be negative.

Tanner and Qi (2005b) have developed a constitutive equation that combines a Newtonian behavior component and a viscoelastic behavior component. The Newtonian behavior component is modeled by a Reiner–Rivlin equation (see Tanner 2000) with a viscosity function depending on the volume fraction of the solid phase. The Reiner–Rivlin equation predicts a zero first normal stress difference and a large negative second normal stress difference for simple shearing. The second normal stress difference can fit experiments exactly by appropriately choosing the model parameter. The viscoelastic behavior component is modeled by a single-mode PTT model, which predicts a positive first normal stress difference and a zero second normal stress difference for simple shearing. Hence the complete response shows a positive first normal stress difference and a negative second normal stress difference. The model has been used to calculate the viscosity and stresses for viscoelastic matrix suspensions. Computed results are in good agreement with experimental measurements reported by Zarraga et al. (2001) and Mall-Gleissle et al. (2002) for shear flow and Le Meins et al. (2003) for elongational flow.

Housiadas and Tanner (2009), following the approach of Greco et al. (2005), have used a perturbation analysis to obtain the analytical solution for the pressure and the velocity field up to $O(\varphi De)$ of a dilute suspension of rigid spheres in a weakly viscoelastic fluid, whereφ is the volume fraction of the spheres and De is the Deborah number of the viscoelastic fluid. The analytical solution was used to calculate the bulk first and second normal stress in simple shear flows and the elongational viscosity. The main results are

$$\langle N_1 \rangle = 2De\left(1 + \frac{5}{2}\varphi\right), \tag{5.61}$$

$$\langle N_2 \rangle = -2De\left(\Xi + \frac{125 + 40(\Xi - 2)}{28}\varphi\right), \tag{5.62}$$

$$\langle \eta_E \rangle = 3\left(1 + \frac{5}{2}\varphi\right) - 3De\left[\Xi - 1 + \frac{5}{28}(8\Xi - 5)\varphi\right], \tag{5.63}$$

where the angular brackets denote the averaged bulk properties, $\Xi = a_G + \zeta$ with a_G being the Giesekus model constant and ζ the slip parameters of the PTT model. The results are written in dimensionless form. The viscosity and stresses are scaled by η_0 (the zero-shear rate viscosity) and $\eta_0\dot{\gamma}$, respectively.

Tanner et al. (2010a) have extended the above results to concentrated regimes by using the Roscoe procedure (Roscoe 1952, also see Phan-Thien and Pham 2000). In concentrated suspensions, some of the fluid is trapped between particles, and hence Roscoe (1952) suggested that the increment of small amount of volume fraction $d\varphi$ results in an effective increase of concentration of $d\varphi/(1 - \varphi/\varphi_m)$, which is called the crowding function, where φ_m is the maximum volume fraction. We use $\langle N_1 \rangle$ as an example to describe the procedure. From Eq. 5.61, one has

$$d\langle N_1 \rangle = \frac{5}{2}(2De)d\varphi. \tag{5.64}$$

Replacing $d\varphi$ by $d\varphi/(1 - \varphi/\varphi_m)$ and noting that $De = N_1/2$ (Housiadas and Tanner 2009), Eq. 5.64 is modified to

$$\frac{d\langle N_1 \rangle}{\langle N_1 \rangle} = \frac{5}{2}\frac{d\varphi}{(1 - \varphi/\varphi_m)}. \tag{5.65}$$

Integrating the equation results in

$$\frac{\langle N_1 \rangle}{\langle N_1 \rangle_0} = \left(1 - \frac{\varphi}{\varphi_m}\right)^{-\frac{5}{2}\varphi_m}, \tag{5.66}$$

where $\langle N_1 \rangle_0$ is the first normal stress difference of the pure matrix fluid.

From what was described above, the Roscoe approach can be briefly summarized as follows:

1. Find the response of a dilute suspension.
2. Assume that a small amount of additional particles are added to the existing suspension, which will enhance the viscosity of the suspension. Replace $d\varphi$ by the effective increase of concentration using a crowding function.
3. By integration, one then finds the properties for finite concentrations.

Tanner et al. (2010b) have applied this method also to suspensions with power-law matrices. For a power-law matrix, $\eta = k(0)\dot{\gamma}^{n-1}$. The viscosity of the suspension is assumed to be

$$\eta = k(\varphi)\dot{\gamma}^{n-1}. \tag{5.67}$$

The effective $k(\varphi)$ of the dilute suspension is given by

$$k(\varphi) = k(0)(1 + ng\varphi), \tag{5.68}$$

where the value of g is obtained by an empirical fitting to Lee and Mear's results (1991):

$$g = \frac{0.383}{n} + 2.117. \tag{5.69}$$

when $n = 1$ and $k(0) = \eta_0$, Eq. 5.68 reduces to the well-known Einstein's result (1906): $\eta = \eta_0(1 + 2.5\varphi)$.

Then one finds the following differential equation:

$$\frac{dk(\varphi)}{d\varphi} = \frac{ngk(\varphi)}{1 - \varphi/\varphi_m}, \tag{5.70}$$

which has the solution:

$$\frac{k(\varphi)}{k(0)} = \left(1 - \frac{\varphi}{\varphi_m}\right)^{-ng\varphi_m}, \tag{5.71}$$

Results of Eq. 5.71 are compared reasonably well with other theories and published experiments.

The work is concerned with suspensions of spheres. The suspension of randomly oriented fibers can be dealt with in much the same way.

Chapter 6
Shrinkage and Warpage

6.1 Introduction

It is known that the dimension of an injection-molded product, as it cools after the molding process, is usually different from the corresponding dimension of the mold cavity. The geometric reduction in the size of the part is referred to as *mold shrinkage*, or *as-molded shrinkage*, or simply *shrinkage*. According to ASTM standards (ASTM D955-08), shrinkage is measured 24–48 h after demolding. *Warpage*, or *warping,* is the distortion induced by the inhomogeneous shrinkage. According to Austin (1991) and Shoemaker (2006), variations in shrinkage can be further classified into three types: (i) shrinkage difference in different directions due to the material anisotropy; (ii) shrinkage variations from region to region in the part due to non-uniform pressure and temperature distributions over the part; (iii) non-uniform shrinkage across the thickness due to the differential cooling on opposing mold faces.

The terms *shrinkage* and *warpage* are sometimes used to describe the dimensional stability of injection-molded part after it exposed to elevated temperature. For example, the baking of a painted car part may cause undesirable changes in size and shape of the part. We call this phenomenon *post-molding shrinkage* and *warpage*. We shall not consider the post-molding shrinkage and warpage in this book. The reader can find discussions on this topic elsewhere (e.g., Fan et al. 2010b, c).

If we consider the dimensional change in volume, we deal with the *volumetric shrinkage*. The volumetric shrinkage can be defined as

$$S_v = \frac{\bar{v}_0(t) - v_r(T_{\text{room}}, P_{\text{atm}})}{\bar{v}_0(t)}, \tag{6.1}$$

where $v_r(T_{\text{room}}, P_{\text{atm}})$ is the specific volume at room temperature under atmospheric pressure, while $\bar{v}_0(t)$ is the gap-wise averaged initial specific volume. When the local cavity pressure has decayed to zero or the material has solidified, whatever occurs first, the value of $\bar{v}_0(t)$ is considered frozen-in, meaning that the

R. Zheng et al., *Injection Molding*, DOI: 10.1007/978-3-642-21263-5_6,

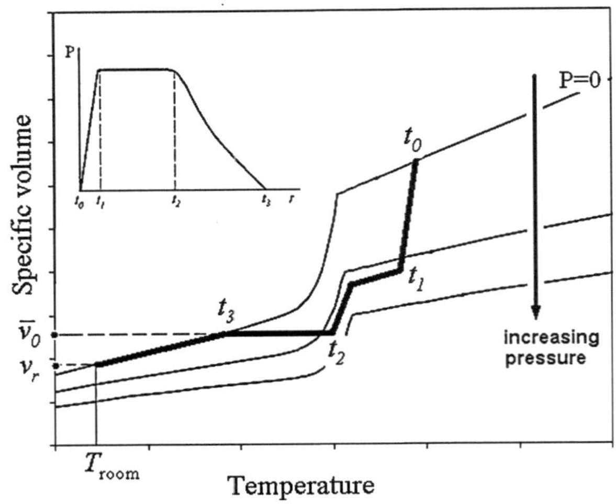

Fig. 6.1 Schematic specific volume evolution with time on the PVT diagram

mass within the control volume under consideration remains unchanged with time, and hence the volumetric shrinkage will approach an asymptotic value. The specific volume depends on the pressure and temperature history and thus the volumetric shrinkage can be calculated using PVT diagrams. The procedure is illustrated in Fig. 6.1, where plotted at the top left is a typical cavity pressure–time profile, and the evolution of specific volume with time is plotted on the PVT program.

Isayev (1987) proposed a slightly different definition for the volumetric shrinkage. He uses the time-average value of the initial specific volume over the packing time, and writes the volumetric shrinkage as

$$S_v = \frac{\frac{1}{t_3-t_1}\int_{t_1}^{t_3} \bar{v}_0(t)dt - v_r(T_{\mathrm{room}}, P_{\mathrm{atm}})}{\frac{1}{t_3-t_1}\int_{t_1}^{t_3} \bar{v}_0(t)dt}. \tag{6.2}$$

Volumetric shrinkage involves the change of dimension in three spatial directions. These changes can be different for each direction. Therefore it is important to consider the dimensional change in length, i.e., the *linear shrinkage*, defined by

$$S_l = \frac{l_0 - l_r}{l_0}, \tag{6.3}$$

where l_0 is the initial length, and l_r is the corresponding length at room temperature under atmospheric pressure. If the shrinkage is isotropic, then the linear shrinkage is related to the volumetric shrinkage by

$$S_l = 1 - (1 - S_v)^{1/3} \approx \frac{1}{3} S_v. \tag{6.4}$$

For anisotropic shrinkage, it is customary to classify the linear shrinkage into *parallel shrinkage* $S^{//}$ (the linear shrinkage in the flow direction), *perpendicular shrinkage* $S^{\perp}$ (the linear in-plane shrinkage perpendicular to the flow direction) and the thickness shrinkage S^h. When $S_v \ll 1$, one has

$$S_v \approx S^{//} + S^{\perp} + S^h. \tag{6.5}$$

No simple relationship between the linear shrinkage components is available. Their values depend on material and processing.

6.2 Mechanical and Thermal Properties of Short-Fiber Composites

Shrinkage and warpage depend on the mechanical and thermal properties of the material. The material properties and performance are determined by its internal structure, and the internal structure can be influenced by processing. This is particularly true in composite materials. The term *composite material* refers to a structure made up of two or more discrete components, which, when combined, can produce superior properties for its intended application. Among many types of composite, short-fiber reinforced polymers evokes the most interest for the injection molding application. A *short-fiber reinforced polymer* is a material containing discontinuous fibers as fillers embedded in a polymer matrix. Glass fiber is the most commonly used filler. While the alignment of the discontinuous fibers in the polymer offers several desired properties, such as high strength, impact resistance or fatigue resistance, the inhomogeneous properties and the orientation effect may lead to undesired shrinkage variations and warpage.

In Chap. 5 we have discussed the flow-induced fiber orientation. In this Chapter we will begin with a particular aspect of material modeling, namely, the modeling of mechanical and thermal properties of the short–fiber composites from the given properties of the individual components and the fiber orientation distributions. A correct evaluation of material properties is a prerequisite for shrinkage and warpage modeling.

6.2.1 Effective Stiffness Tensor of Unidirectional Composites

For infinitesimal deformations considered in injection molding simulations, the total strain $\varepsilon_{ij}^{(\mathrm{Total})}$ is regarded as the sum of elastic strain ε_{ij} and the thermal strain $\varepsilon_{ij}^{(\mathrm{th})}$,

$$\varepsilon_{ij}^{(\mathrm{Total})} = \varepsilon_{ij} + \varepsilon_{ij}^{(\mathrm{th})}. \tag{6.6}$$

In the classical theory of elasticity, the stress tensor σ_{ij} is linearly dependent on the elastic strain tensor, i.e.,

$$\sigma_{ij} = C_{ijkl}\varepsilon_{kl} = C_{ijkl}\left(\varepsilon_{kl}^{(\text{total})} - \varepsilon_{kl}^{(\text{th})}\right), \tag{6.7}$$

where C_{ijkl} is a fourth order tensor of elastic constant, called the stiffness tensor.

The elastic strain can be expressed as a function of the stress by

$$\varepsilon_{ij} = S_{ijkl}\sigma_{kl}, \tag{6.8}$$

where $S_{ijkl} = C_{ijkl}^{-1}$ is called the elastic compliance.

For the fiber-reinforced polymer materials, the C_{ijkl} represents the effective stiffness tensor of the composite, which links the average strain to the average stress by

$$\bar{\sigma}_{ij} = C_{ijkl}\bar{\varepsilon}_{kl}. \tag{6.9}$$

Here the average of any quantity means its integral over a specified region divided by the volume of the region. Consider a volume V which is large enough to contain many fibers but small enough so that the macroscopic variables hardly change on the scale $V^{1/3}$. The effective stress seen from a macroscopic level is simply the volume-averaged stress:

$$\bar{\sigma}_{ij} = \frac{1}{V}\int\limits_{V} \sigma_{ij} dV = \frac{1}{V}\int\limits_{V^f} \sigma_{ij} dV + \frac{1}{V}\int\limits_{V^m} \sigma_{ij} dV, \tag{6.10}$$

where V^f and V^m are the volumes occupied by the inclusions (fibers) and the matrix (polymer) respectively. Here, and in what follows, we shall use superscript f to denote fiber and m to denote matrix.

If we define the average fiber and matrix stresses as

$$\bar{\sigma}_{ij}^f = \frac{1}{V^f}\int\limits_{V^f} \sigma_{ij} dV, \quad \bar{\sigma}_{ij}^m = \frac{1}{V^m}\int\limits_{V^m} \sigma_{ij} dV, \tag{6.11}$$

and let $\varphi^f = V^f/V$ be the volume fraction of fibers, and $\varphi^m = V^m/V$ be the volume fraction of the matrix, then from Eq. 6.10 we have,

$$\bar{\sigma}_{ij} = \varphi^f \bar{\sigma}_{ij}^f + \varphi^m \bar{\sigma}_{ij}^m. \tag{6.12}$$

The average strains $\bar{\varepsilon}_{ij}$, $\bar{\varepsilon}_{ij}^f$ and $\bar{\varepsilon}_{ij}^m$ are defined similarly, leading to

$$\bar{\varepsilon}_{ij} = \varphi^f \bar{\varepsilon}_{ij}^f + \varphi^m \bar{\varepsilon}_{ij}^m. \tag{6.13}$$

The stress–strain relations in the fiber and matrix phases are

$$\bar{\sigma}_{ij}^f = C_{ijkl}^f \bar{\varepsilon}_{kl}^f, \tag{6.14}$$

and

$$\bar{\sigma}^m_{ij} = C^m_{ijkl}\,\bar{\varepsilon}^m_{kl}, \tag{6.15}$$

respectively. Substituting Eqs. 6.14 and 6.15 into 6.12 gives

$$\bar{\sigma}_{ij} = \varphi^f C^f_{ijkl}\,\bar{\varepsilon}^f_{kl} + \varphi^m C^m_{ijkl}\,\bar{\varepsilon}^m_{kl}. \tag{6.16}$$

Hill (1963) suggested that there is a unique dependence of the fiber-phase average strain, $\bar{\varepsilon}^f_{ij}$, on the overall average strain of the composite, $\bar{\varepsilon}_{ij}$, given as

$$\bar{\varepsilon}^f_{ij} = A_{ijkl}\,\bar{\varepsilon}_{kl}, \tag{6.17}$$

where the fourth-order tensor A_{ijkl} is called the strain-concentration tensor. The matrix-phase strain $\bar{\varepsilon}^m_{ij}$ can also be related to $\bar{\varepsilon}_{ij}$ by substituting Eq. 6.17 into 6.13 to obtain

$$\varphi^m\,\bar{\varepsilon}^m_{ij} = (I_{ijkl} - \varphi^f A_{ijkl})\bar{\varepsilon}_{kl}, \tag{6.18}$$

where $I_{ijkl} = (1/2)(\delta_{ik}\delta_{jl} + \delta_{il}\delta_{jk})$ is the fourth-order unit tensor. Substituting Eqs. 6.17 and 6.18 into Eq. 6.16 gives

$$\bar{\sigma}_{ij} = C_{ijkl}\bar{\varepsilon}_{kl}, \tag{6.19}$$

where

$$C_{ijkl} = C^m_{ijkl} + \varphi^f\left(C^f_{ijmn} - C^m_{ijmn}\right)A_{mnkl}. \tag{6.20}$$

Equation 6.20 is the required equation for the effective stiffness tensor C_{ijkl}. Since C^m_{ijkl}, C^f_{ijkl} and φ^f are all known, one only needs to find the strain-concentration tensor A_{ijkl}. Different expressions of A_{ijkl} represent different models. Many models have been reviewed by Tucker and Liang (1999). They recommend the Mori–Tanaka model as the best choice for injection molded composites. The model was proposed by Mori and Tanaka (1973) and has later been described by Benveniste (1987) and Christensen (1990) in a simpler direct way. The Mori–Tanaka strain-concentration tensor is given by

$$A_{mnkl} = \left[I_{mnkl} + \left(1 - \varphi^f\right)E_{mnrs}\left(C^m_{rspq}\right)^{-1}\left(C^f_{pqkl} - C^m_{pqkl}\right)\right]^{-1}, \tag{6.21}$$

where E_{ijkl} is known as the Eshelby tensor (Eshelby 1957). Consider an ellipsoid inclusion whose three principal axes are a_1, a_2 and a_3, respectively, and a_1 is its major radius. A rectangular Cartesian coordinate system $[x_i(i = 1,\ 2,\ 3)]$ is introduced such that the origin coincides the center of the ellipsoid and axes x_i aligned with the principal axes a_i. For a generic anisotropic material the Eshelby tensor is given by (Mura 1987)

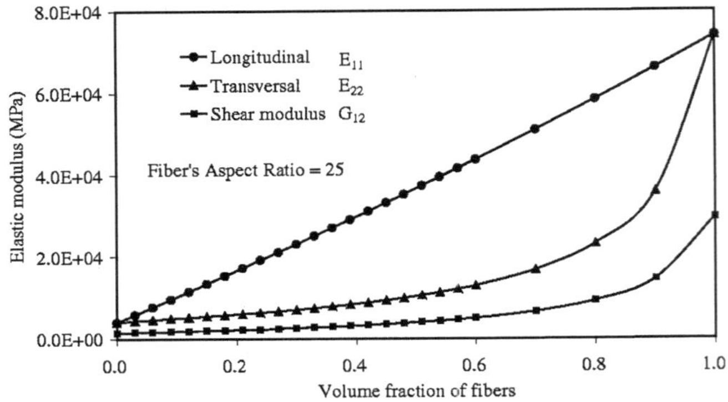

Fig. 6.2 Numerical results based on the Mori–Tanaka model for the elastic moduli E_{11}, E_{22} and G_{12} as a function of volume fraction of fibers

$$E_{ijkl} = \frac{1}{8\pi} C^m_{pqkl} \int_{-1}^{1} d\bar{\zeta}_1 \int_{0}^{2\pi} \left[G_{ipjq}(\bar{\xi}) + G_{jpiq}(\bar{\xi}) \right] d\theta, \tag{6.22}$$

in which

$$G_{ijkl}(\bar{\xi}) = \bar{\xi}_k \bar{\xi}_l N_{ij}(\bar{\xi}) / \Omega(\bar{\xi}), \text{ with } \bar{\xi}_k = \bar{\zeta}_k / a_k, \tag{6.23}$$

where $\bar{\zeta}_1$ is an integration variable in $[-1, 1]$,

$$\bar{\zeta}_2 = (1 - \bar{\zeta}_1^2)^{1/2} \cos\theta, \tag{6.24}$$

$$\bar{\zeta}_3 = (1 - \bar{\zeta}_1^2)^{1/2} \sin\theta, \tag{6.25}$$

$$\Omega(\bar{\xi}) = \varepsilon_{mnl} K_{m1} K_{n1} K_{l1}, \tag{6.26}$$

$$N_{ij}(\bar{\xi}) = \frac{1}{2} \varepsilon_{ikl} \varepsilon_{jpq} K_{kp} K_{lq}, \tag{6.27}$$

$$K_{ik} = C^m_{ijkl} \bar{\xi}_j \bar{\xi}_l, \tag{6.28}$$

and ε_{ijk} is the permutation tensor.

For a special case that the properties of the matrix are isotropic matrix case, closed form expressions of the Eshelby tensor have been obtained by Tandon and Weng (1984). In general, explicit evaluation of E_{ijkl} is a difficult problem, but it can be calculated numerically (Gavazzi and Lagoudas 1990).

As an example, numerical results of elastic properties (E_{11}, E_{22} and G_{12}) for a glass fiber-reinforced polymer are shown in Fig. 6.2. The component properties used are:

Matrix	**Fiber**
$E^{(m)} = 4,000\text{MPa}$	$E^{(f)} = 74,000\,\text{MPa}$
$\upsilon^{(m)} = 0.36$	$\upsilon^{(f)} = 0.25$

where $E^{(m)}$ and $\upsilon^{(m)}$ denote the Young's modulus and the Poisson's ratio of the matrix, respectively, $E^{(f)}$ and $\upsilon^{(f)}$are the Young's modulus and the Poisson's ratio of the glass fibers.

6.2.2 *Effective Thermal Expansion Coefficients of Unidirectional Composites*

The effective thermal expansion coefficient of the composite, α_{ij}, related to the average strain tensor, $\bar{\varepsilon}_{ij}$, in the following equation:

$$\bar{\varepsilon}_{ij} = S_{ijkl}\,\bar{\sigma}_{kl} + \alpha_{ij}\Delta T, \tag{6.29}$$

where ΔT is the temperature rise, S_{ijkl} the effective elastic compliance that can be obtained from the Mori–Tanaka model by inversion.

Rosen and Hashin (1970) have obtained the relation between the effective thermal expansion coefficient and the properties of the components:

$$\alpha_{ij} = \bar{\alpha}_{ij} + P_{klmn}\left(S_{mnij} - \bar{S}_{mnij}\right)\left(\alpha^f_{kl} - \alpha^m_{kl}\right), \tag{6.30}$$

where, again, superscripts f and m refer to fiber and matrix, respectively, and terms with an overbar refer to volume-average composite properties, i.e.,

$$\bar{\alpha}_{ij} = \varphi^f \alpha^f_{ij} + \left(1 - \varphi^f\right)\alpha^m_{ij}, \tag{6.31}$$

and

$$\bar{S}_{mnij} = \varphi^f S^f_{mnij} + (1 - \varphi^f)S^m_{mnij}. \tag{6.32}$$

The tensor P_{klmn} is determined from the following equation:

$$P_{klmn}\left(S^f_{mnrs} - S^m_{mnrs}\right) = I_{klrs}. \tag{6.33}$$

Figure 6.3 shows the longitudinal and the transverse coefficients of thermal expansion (the CTEs) predicted using the Rosen–Hashin model. The component properties of the composite used are:

Matrix	**Fiber**
$E^{(m)} = 4,000\,\text{MPa}$	$E^{(f)} = 74,000\,\text{MPa}$
$\upsilon^{(m)} = 0.36$	$\upsilon^{(f)} = 0.25$
$\alpha^{(m)} = 6.0 \times 10^{-5}K^{-1}$	$\alpha^{(f)} = 5.0 \times 10^{-6}K^{-1}$

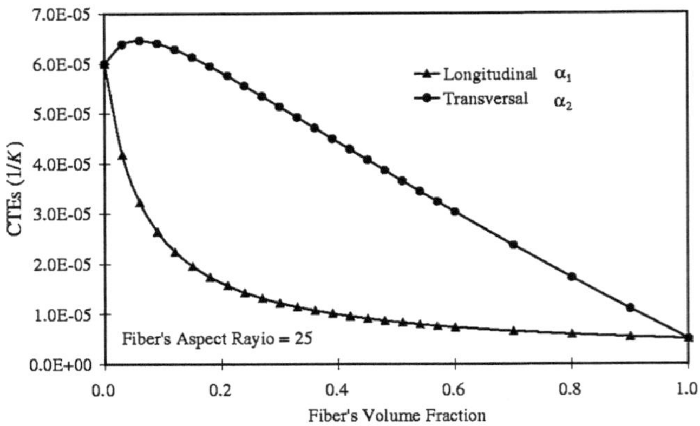

Fig. 6.3 Numerical results based on the Rosen–Hashin model for the coefficients of thermal expansion as a function of volume fraction of fibers

6.2.3 Orientation Averaging

In reality, the fibers in a short-fiber reinforced composite are rarely completely aligned, so that a further step of calculations is required to account for the effect of the distribution of fiber orientation on the actual properties of the composite. The procedure of averaging properties over all directions by writing with the orientation distribution is termed *orientation averaging*. The orientation averaging can be formulated in terms of the orientation tensor a_{ij} and a_{ijkl}.

Consider a subunit that comprised of fully aligned short fibers. The subunit is said to be *transversely isotropic*, if it is isotropic in all planes perpendicular to the fiber direction, and the fiber axis is the axis of asymmetry. For a transversely isotropic unidirectional stiffness tensor C_{ijkl}, orientation averaging can be done to give the composite stiffness, $\langle C_{ijkl} \rangle$, as follows (Advani and Tucker 1987):

$$\begin{aligned}\langle C_{ijkl} \rangle = & B_1 a_{ijkl} + B_2(a_{ij}\delta_{kl} + a_{kl}\delta_{ij}) + B_3(a_{ik}\delta_{jl} + a_{il}\delta_{jk} + a_{jl}\delta_{ik} + a_{jk}\delta_{il}) \\ & + B_4(\delta_{ij}\delta_{kl}) + B_5(\delta_{ik}\delta_{jl} + \delta_{il}\delta_{jk}).\end{aligned} \tag{6.34}$$

The constants B_i are invariants of the tensor C_{ijkl}, given by

$$B_1 = C_{1111} + C_{2222} - 2C_{1122} - 4C_{1212}, \tag{6.35}$$

$$B_2 = C_{1122} - C_{2233} \tag{6.36}$$

$$B_3 = C_{1212} + \frac{1}{2}(C_{2233} - 2C_{2222}), \tag{6.37}$$

$$B_4 = C_{2233}, \tag{6.38}$$

$$B_5 = \frac{1}{2}(C_{2222} - C_{2233}). \tag{6.39}$$

In general, $\langle C_{ijkl} \rangle$ is no longer transversely isotropic.

Lin et al. (2004) adopted the same approach but they extended the five-constant equation to a nine-constant equation, which relaxes the transversely isotropic assumption of the properties for the composite of aligned inclusions, and can be use for more general orthotropic properties. Eduljee et al. (1994) presented an orientation averaging approach capable of distinguishing between dispersed and aggregated microstructures in short-fiber composites.

The orientation averaging method can also be applied to thermal expansion coefficients. To account for the effect of fiber orientation distribution, one writes the thermoelastic constitutive equation for the unidirectional composite as

$$\sigma_{ij} = C_{ijkl}\left(\varepsilon_{kl}^{\text{total}} - \alpha_{kl}\Delta T\right), \tag{6.40}$$

then the average stress is, assuming that fibers in all orientation experience the same strain and the same temperature difference ΔT,

$$\langle \sigma_{ij} \rangle = \langle C_{ijkl} \rangle \varepsilon_{kl}^{\text{total}} - \langle C_{ijmn}\alpha_{mn} \rangle \Delta T. \tag{6.41}$$

The thermal expansion gives the strains where $\langle \sigma_{ij} \rangle = 0$, which is

$$\varepsilon_{kl}^{\text{total}} = \langle C_{ijkl} \rangle^{-1} \langle C_{ijmn}\alpha_{mn} \rangle \Delta T. \tag{6.42}$$

So we find that

$$\langle \alpha_{kl} \rangle = \langle C_{ijkl} \rangle^{-1} \langle C_{ijmn}\alpha_{mn} \rangle, \tag{6.43}$$

where the average of the fourth order stiffness tensor is given in Eq. 6.34 above. The other orientation average for the second order tensor is given by Bay (1991):

$$\begin{aligned}\langle C_{ijmn}\alpha_{mn} \rangle = &[(C_{1111} - C_{1122})\alpha_{11} + (2C_{1122} - C_{2222} - C_{2233})\alpha_{22}]a_{ij} \\ &+ [C_{2211}\alpha_{11} + (C_{2222} + C_{2233})\alpha_{22}]\delta_{ij}\end{aligned} \tag{6.44}$$

In what follows in this chapter, we shall drop the angular brackets, and simply use C_{ijkl} and a_{ij} to represent the orientation-averaged properties of composites.

6.3 Thermally and Pressure-Induced Stresses

6.3.1 Stress Development

In the injection molding process, frozen layers develop from the outer surface toward the core; each layer solidifies at a different time under different pressures. While all layers undergo the tendency of thermal contraction that generates

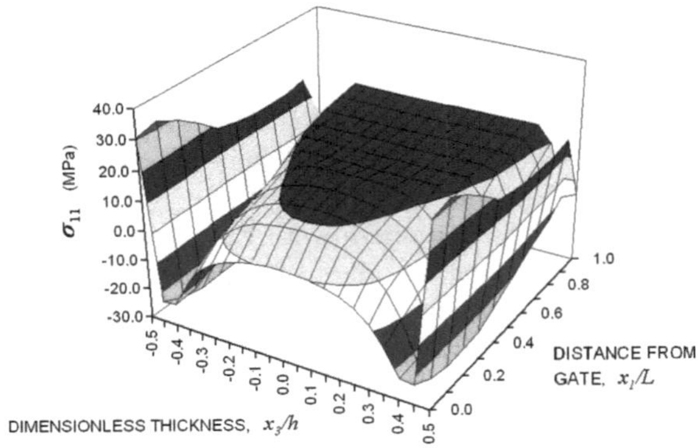

Fig. 6.4 Thermally and pressure-induced stress (σ_{11}) distribution (From Zheng et al. (1999), with permission from Elsevier)

in-plane tensile stresses due to constraints of the mold, the layers solidified under high pressure would tend to build up compressive stresses. The combination of the pressure and thermal effects results in the typical stress profile in an injection molded plate as shown in Fig. 6.4 (Zheng et al. 1999), where one sees tensile stresses at the mold-polymer interface, followed by compressive stresses in a region inward, and tensile stresses again in the core region.

The development of the stresses can be illustrated at sequential time instants by means of numerical results. Figure 6.5 shows a cavity pressure evolution profile. Figure 6.6 shows the gap-wise in-plane stress profiles of σ_{11} at successive times. Five typical time instants are chosen to display the results. They are: $t_1 = 0.56$ s when the location has just been filled; $t_2 = 0.86$ s at the end of the filling stage; $t_3 = 1.68$ s when the pressure reaches the peak value; $t_4 = 10.0$ s after the core is solidified (Complete solidification occurred at $t = 8.7$ s); $t_5 = 35.9$ s just before demolding.

As can be seen, before complete solidification the stresses in the molten core equal the negative of the fluid pressure, while the stresses in the frozen skin layer increase as consequence of the constrained cooling. After complete solidification across the thickness, the stresses throughout the whole thickness increase due to the decrease in both pressure and temperature. The final profile at the time $t_5 = 35.9$ s shows that considerable tensile stresses have been built up in the surface layer, followed by a compressive stress region further inward, then an increase which causes the stresses to become tensile again in the core. Clearly, the existence of the relatively high tensile stresses in the skin and core regions is associated with the relatively low pressure under which the skin and the core have frozen, while the existence of the minima in the intermediate region is due to the fact that the material in this region has frozen at the maximum pressure. This pattern of stress profile is typical for injection-molded parts, though exceptions indeed exist.

At each time step, the integral of stresses over the thickness is balanced with the external forces applied to the part by the mold walls. When the part is ejected from the mold, the external forces applied to the part surface are suddenly removed. The product can thus deform according to force balance. Some residual stresses can remain within the part. The equilibrium of the residual stresses determines the

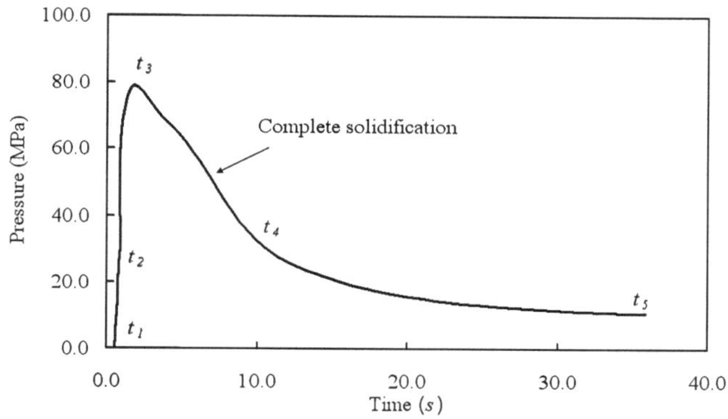

Fig. 6.5 A pressure evolution profile

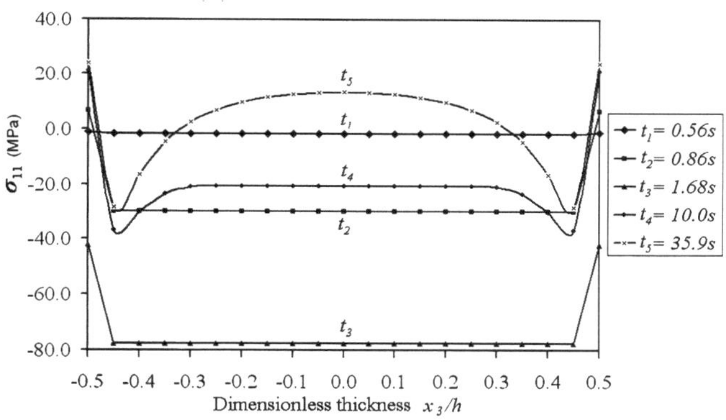

Fig. 6.6 Gap-wise profiles of in-plane σ_{11}-stress at successive times (From Zheng et al. (1999), with permission from Elsevier)

dimensions of the part. The thermally and pressure-induced stress profile can be asymmetric about the mid-surface due to asymmetrical cooling. The skin region will be thicker on the cold side than on the hot side, and this will shift the core region toward the hot side. As a result, the bending moment of the stresses may cause the part to warp toward the hotter face after demolding. However, exceptions exist, depending on the pressure patterns, as discussed by Boitout et al. (1995).

6.3.2 Viscous-Elastic Model and Viscoelastic Model

In the calculation of the thermally and pressure-induced stresses, the material behavior can be described by either a *viscous-elastic* model (see, for example, Titomanlio et al. 1987; Jansen 1994; Boitout et al. 1995) or a *viscoelastic* model (see for example, Baaijens 1991; Santhanam et al. 1991; Kabanemi and Crochet 1992; Bushko and Stokes 1995; Zoetelief et al. 1996; Caspers 1996; Zheng et al. 1999; Mlekusch 2001; Kamal et al. 2002; Kim et al. 2007).

The *viscous-elastic* model, also called the *thermo-elastic* model, was proposed by Struik (1990). The model assumes that the thermal elastic parameters are constant and independent of temperature below as well as above the solidification temperature T_s, and that the only discontinuous changes occur at T_s. Usually, the material is assumed to be liquid and to sustain no stresses above T_s, while below T_s the material is assumed elastic and able to sustain stresses. Under this assumption we have

$$\sigma_{ij} = \begin{cases} 0 & T \geq T_s \\ C^{(e)}_{ijkl}(\varepsilon_{kl} - \alpha_{kl}\Delta T) & T < T_s \end{cases}, \quad (6.45)$$

The *viscoelastic* model used in the calculation is a linear viscoelastic model with the assumption of thermorheological simplicity, given by

$$\sigma_{ij} = \int_0^t C_{ijkl}(\xi(t) - \xi(t'))\left(\frac{\partial \varepsilon_{kl}}{\partial t'} - \alpha_{kl}\frac{\partial T}{\partial t'}\right)dt', \quad (6.46)$$

with

$$\xi(t) = \int_0^t \frac{dt'}{a_T}, \quad (6.47)$$

where $\xi(t)$ is the pseudo-time scale, and a_T is the time–temperature shift factor. Here we write the memory function in terms of the fourth-order stiffness tensor $C_{ijkl}(t)$, in order to model linear viscoelastic response of fiber-reinforced composites. For isotropic materials,

$$C_{ijkl}(t) = \frac{2\nu G(t)}{1 - 2\nu}\delta_{ij}\delta_{kl} + G(t)\left(\delta_{ik}\delta_{jl} + \delta_{il}\delta_{jk}\right), \quad (6.48)$$

where ν is Poisson's ratio and $G(t)$ is the shear relaxation modulus given by

$$G(t) = \sum_{i=1}^{N} G_i \exp(-t/\lambda_i), \quad (6.49)$$

in which the set of N pairs of (G_i, λ_i) is called relaxation spectrum. For isotropic materials, we also have

$$\alpha_{kl} = \frac{1}{3}\alpha_V \delta_{kl}, \quad (6.50)$$

where α_V is the volumetric thermal expansion coefficient. Substituting (6.48) and (6.50) into Eq. 6.46 gives

$$\sigma_{ij} = -P^h \delta_{ij} + 2 \int_0^t G(\xi(t) - \xi(t')) \frac{\partial \varepsilon_{ij}^d}{\partial t'} dt', \tag{6.51}$$

where

$$\varepsilon_{ij}^d = \varepsilon_{ij} - \frac{1}{3} \varepsilon_{kk} \delta_{ij}, \tag{6.52}$$

and

$$P^h = \int_0^t K_B \left(\alpha_V \frac{\partial T}{\partial t'} - \frac{\partial \varepsilon_{kk}}{\partial t'} \right) dt', \tag{6.53}$$

with K_B being the bulk modulus

$$K_B = \frac{2G(1+v)}{3(1-2v)}. \tag{6.54}$$

The relaxation of the bulk modulus has been neglected here, while it was considered by Ghoneim and Hieber (1997).

While in the mold, the material is constrained by the mold in all in-plane directions, so that tensile stresses will develop when the material is cooled down. The stresses will relax a certain amount depending on the temperature history during cooling. This stress relaxation phenomenon in injection molding is known as the effect of in-mold constraints. The effect can only be predicted by the viscoelastic model but not the viscous-elastic model. However, if the Deborah number (defined here as the ratio between the relaxation time of the material and the cooling time) is much larger than one, the viscous-elastic model is able to give a good qualitative description of the thermally and pressure-induced stresses (Zoetelief et al. 1996), and provides considerable simplification. The critical point in using the model is the choice of the solidification temperature T_s. With the viscous-elastic model, material properties change discontinuously at T_s, but T_s is not a clearly defined fundamental physical property. The temperature can be different from the previous mentioned "no-flow temperature" (see Sect. 3.2). The concept of "solidification" here does not just mean that the melt velocity effectively becomes zero. Rather, it means that the material can be considered as an elastic solid. Some authors (e.g., Zoetelief et al. 1996; Kamal et al. 2002) have used the transition temperature T_{trans} (or, T_g for amorphous materials) determined from the DSC measurements as the solidification temperature. The temperature T_{trans} is about 10–30°C below T_{noflow}. Another possibility is to relate the solidification temperature to the stress relaxation, and define T_s as the temperature below which no more stress relaxation occurs on a time scale comparable to the processing time (see Jansen 1994). The solidification temperature so defined is about 10–20°C below T_{trans}. A different selection for the solidification temperature would certainly lead to different results, especially the predicted location of the compressive stress layer.

On the other hand, the viscoelastic model does not require any solidification criterion in principle. Although one may use a solidification temperature to distinguish the "liquid" and "solid" phases for convenience, there is no discontinuous change in material properties at this point. The material naturally becomes stiffer as the relaxation times increase with decreasing temperature. However, the viscoelastic model poses considerable complexity in material characterization. Although the time–temperature superposition principle has been widely used to simplify experimental procedures and numerical calculations for thermorheologically simple polymers, many polymers are unfortunately not thermorheologically simple. Thermorheological simplicity demands that all the molecular mechanisms involved in the relaxation process have the same temperature dependence. Dutta and Edward (1997) has shown that semicrystalline materials such as the isotactic polypropylene (iPP) exhibit different temperature dependencies associated with the crystalline phase-related relaxation and the amorphous phase-related relaxation, respectively, and such thermorheologically complex materials are not reducible by simple superposition principles.

6.3.3 Assumptions and Boundary Conditions

For thin-walled injection-molded parts, the calculation of thermally and pressure-induced stresses can be simplified considerably by using the thin-film approximation. This enables us to apply the following assumptions:

1. With respect to the local coordinates in which the x_3-direction is normal to the local midplane, the shear strains $\varepsilon_{13} = \varepsilon_{23} = 0$.
2. The normal stress σ_{33} is constant across the thickness, and as long as $\sigma_{33} < 0$, the material sticks to the mould walls.
3. Before ejection, the part is fully constrained within the plane of the part such that the only non-zero component of strain is ε_{33}.

The following different cases are considered for the local boundary conditions in terms of the normal stress σ_{33}:

Case 1 The part is in the mold and there is coexistence of a solid layer and molten core. In this case, the normal stress σ_{33} equals the opposite of the fluid pressure:

$$\sigma_{33} = -P. \tag{6.55}$$

The pressure is calculated from flow analyses of filling and packing.

Case 2 The part is still in the mold and the material has solidified throughout the thickness, the material may either stick to the mould walls or detach from the walls. If it still contacts the walls, then σ_{33} should be determined by writing that the integral of the ε_{33} over the thickness equals the variation of cavity thickness. If the wall deformation is neglected, one has

$$\int_{-h/2}^{h/2} \varepsilon_{33} dx_3 = 0. \tag{6.56}$$

If the polymer detaches from the wall, then

$$\sigma_{33} = 0. \tag{6.57}$$

With the above values of σ_{33}, the in-plane stress components can then be deduced.

Case 3 The part is ejected, no outside loads or constraints are applied to the part, thus the boundary condition is a zero surface traction:

$$\sigma_{ij} n_j = 0, \tag{6.58}$$

where n_j is the normal outward unit vector on the boundary.

The above assumptions and boundary conditions have been implemented in a 2.5D simulation of injection molding (Zheng et al. 1999, see also Chap. 8 for details of the 2.5D finite element method). Residual stress calculations for 3D elements have recently been tackled by Fan et al. (2010d).

6.4 Displacement Calculation

When the part is fully constrained in the mold, no calculation of the displacements is required. After the part is ejected, however, all strain components may be non-zero, and can be evaluated from the following elastic stress–strain relation:

$$\sigma_{ij} = C_{ijkl}(\varepsilon_{kl} - \varepsilon_{kl}^{\text{th}}) + \sigma_{ij}^{I} \tag{6.59}$$

where $\varepsilon_{kl}^{\text{th}}$ is the thermal strain due to the free quench during the subsequent cooling from the ejection temperature to the room temperature. σ_{ij}^{I} is the initial stress, i.e., the thermally and pressure-induced residual stresses generated during the cooling inside the mold (Frozen-in flow-induced residual stresses may also be included).

Alternatively, for simplicity, one can merge the free-quench thermal strains into the initial stresses term and perform an isothermal structural analysis at the room temperature. In this case, Eq. 6.59 is rewritten as

$$\sigma_{ij} = C_{ijkl}\, \varepsilon_{kl} + \sigma_{ij}^{0}. \tag{6.60}$$

Here the single term σ_{ij}^{0} replaces $(\sigma_{ij}^{I} - C_{ijkl}\, \varepsilon_{kl}^{th})$ in Eq. 6.59. If Eq. 6.60 is to be used, the calculation procedure for the in-mold thermally and pressure-induced stress has to be continued to the room temperature. If the material has fully frozen

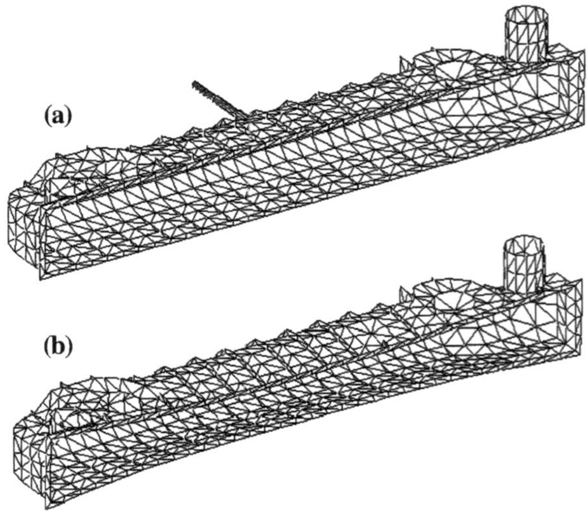

Fig. 6.7 Predicted deformation of an injection molded part: **a** original, **b** deformed. Displacements are amplified by a factor of 2 (Reproduced from Zheng et al. (1996), with permission from Sage Science Press)

and behaves like an elastic solid upon ejection, both methods should yield the same results. Otherwise, for example, if the part is ejected with a molten core, Eq. 6.60 is considered to be only an approximate treatment.

Equations 6.59 or 6.60 with given boundary conditions can be solved for the displacement field using numerical methods such as finite element methods. Figure 6.7 gives an example of predicted deformation of an injection-molded part (Zheng et al. 1996).

6.5 Empirical Approach

As an alternative to the above-mentioned residual-stress based approach is a method based on multi-variable regression technique. For example, Walsh (1993) proposed expressions for the parallel shrinkage and the perpendicular shrinkage as follows:

$$S^{//} = a_1 M_1 + a_2 M_2 + a_3 M_3 + a_4 M_4^{//} + a_5, \tag{6.61}$$

$$S^{\perp} = b_1 M_1 + b_2 M_2 + b_3 M_3 + b_4 M_4^{\perp} + b_5, \tag{6.62}$$

where $S^{//}$ and $S^{\perp}$ are the parallel and perpendicular shrinkages, M_1 to $M_4^{//}$ (and $M_4^{\perp}$) represent the effects of processing and calculated from filling and packing analyses. The measures of effects include: (i) volumetric shrinkage, (ii) degree of crystallinity; (iii) frozen-in stress relaxation as a measure of the effect of in-mold constraints, and (iv) material orientation. The coefficients a_1 to a_5 and b_1 to b_5 are determined by fitting the equations to experimental data.

The samples used to collect experimental information are 200 mm × 40 mm rectangular plates with different thicknesses from 1.7 to 5 mm. Each plate is molded with a fine rectangular grid pattern etched on one surface for shrinkage measurement. Typically, a total of 28 sets processing conditions are used to cover as broad a molding range as possible. Processing sets are created by combinations of varying sample thickness, melt temperature, mold temperature, injection speed, packing pressure, hold time, and cooling time. After molding, the samples remain for an annealing period (typically 10 days) in a controlled environment atmosphere, before they are used for shrinkage measurement. Some examples of measured shrinkage data for several typical polymers can be found in Kennedy and Zheng (2006).

The coefficients a_k and b_k ($k = 1,\ldots, 5$) are assumed to depend on material only. After obtained the coefficients with the associated material, they are input to filling and packing analyses to calculate the parallel and perpendicular shrinkages. The parallel and perpendicular shrinkages are then treated as the thermal strains (or converted to initial stresses) and passed to the structural analysis to calculate the shrinkage and warpage.

Another approach described by Kennedy and Zheng (2002) is to introduce some correction factors to the residual-stress based model, and use the experimental shrinkage data to determine the correction factors. The semi-empirical method is known as the CRIMS (Corrected Residual In-Mold Stress) approach.

6.6 Corner Deformation

Injection molded parts with corners (i.e., the local "L" or "C" geometry) often show inward corner warpage (Ammar et al. 2003; Bakharev et al. 2005). This phenomenon is believed to be caused by the following two effects:

The first is the asymmetric cooling effect. In the corner region, heat transfer during cooling is slower inside the corner than outside. Thus, as has been discussed by Boitout et al. (1995), in the case that the gate freezes before the total thickness is totally solidified, the corner will warp toward the hotter side.

The second effect is known as the "spring forward effect" (see, also, Albert and Fernlund 2002). Let us consider a cross-section of a round corner as shown in Fig. 6.8. The shrinkage in the direction of the length of the arc (the in-plane shrinkage) is $S_l = -dl/l$; the strain component along the radial direction (the thickness shrinkage) is $S_r = -dr/r$. Note that $l = r\theta$. By taking logarithm in both side of the equation and then differentiating it, one obtains $dl/l = dr/r + d\theta/\theta$, i.e.,

$$\frac{d\theta}{\theta} = S_r - S_l. \tag{6.63}$$

As the resulting angle deformation is independent of the radius r, the same result can be obtained for sharp corner ($r \to 0$) as well. Equation 6.63 indicates that the change in the corner angle (the internal corner angle is $\pi - \theta$) is due to the

Fig. 6.8 Schematic representation of a round corner

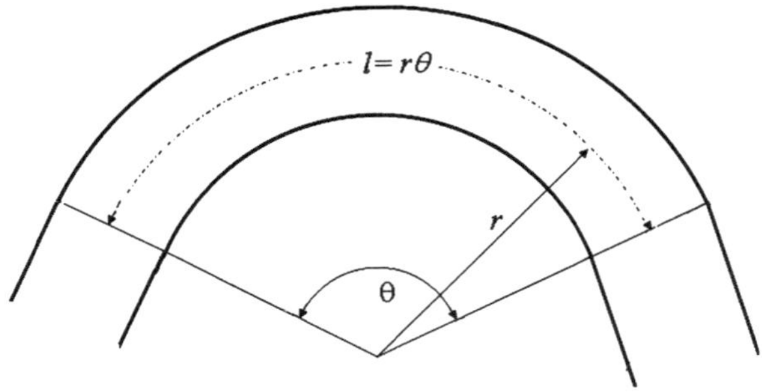

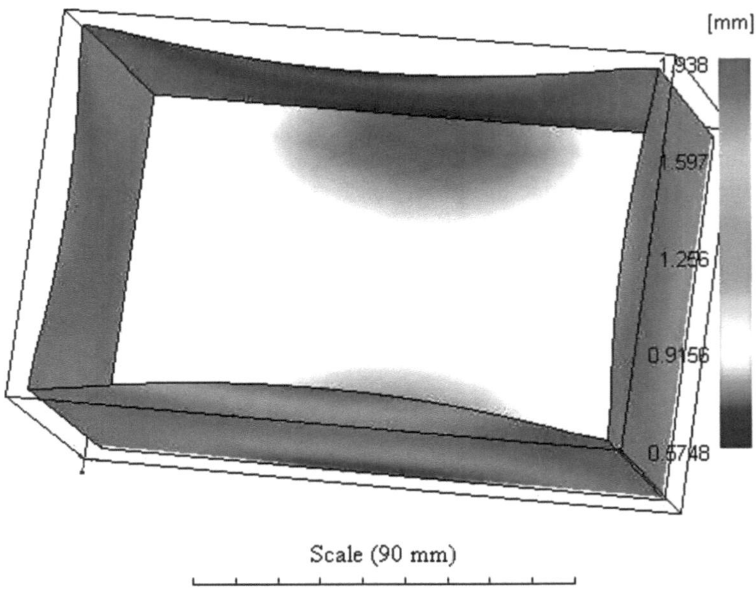

Fig. 6.9 Warpage simulation result of a box-like part (From Bakharev et al. (2005), with permission from Society of Plastics Engineers Inc.)

differences between the thickness shrinkage and the in-plane shrinkage. Usually, the thickness shrinkage is higher than the in-plane shrinkage, so we often see a reduction of the internal corner angle.

For warpage simulations using shell elements (Chap. 8), the thickness shrinkage is not involved in the structural analyses, and the spring forward effect must therefore be imposed. Figure 6.9 displays a predicted deflection of a box-like part (Bakharev et al. 2005). The concave warpage of the box is due to the corner effect.

Chapter 7
Mold Cooling

7.1 Mold Cooling System

In injection molding, the mold has two functions: (i) to form the shape of the part to be manufactured, and (ii) to extract heat from the material to solidify the part as quickly as possible. For performing the second function, the mold has a cooling system within it. The basic cooling system consists of the cooling lines in the form of circular holes drilled in the mold so that the coolant flowing through the cooling lines will extract the heat out from the hot polymer melt. The location of cooling lines depends on the part geometry, cavity configuration, and the location of ejection pins and moving components of the mold. Figure 7.1 illustrates the cooling lines and their relation to the part. In the figure the cooling system consists of two circuits, one (Circuit 1) in the fixed half of the mold, and the other (circuit 2) in the moving half. This example has been used by Zheng et al. (1996) in a study aimed at prediction of warpage.

The mold may consist of regions where circuits are not feasible. Additional cooling channels such as baffles, bubblers, or thermal pins may be used to divert the coolant flow into these regions. Shoemaker (2006) has shown some examples.

In the production of injection-molded thermoplastic parts, about three-fourths of the total cycle time is spent in cooling the hot polymer sufficiently. An efficient cooling circuit design can minimize the cooling time. Mold temperature has significant effects on the polymer flow and solidification behavior, and improper cooling is one of the major causes of defects in molded parts, such as warpage. Therefore, in injection molding simulations, mold-cooling analysis is essential for two purposes: one is to be used as a numerical tool for cooling system design; and the other is to provide thermal boundary conditions for filling and packing analyses.

7.2 Transient Heat Transfer in Mold

The temperature field in the mold is governed by the three-dimensional transient heat conduction equation with constant properties. The equation is

R. Zheng et al., *Injection Molding*, DOI: 10.1007/978-3-642-21263-5_7,

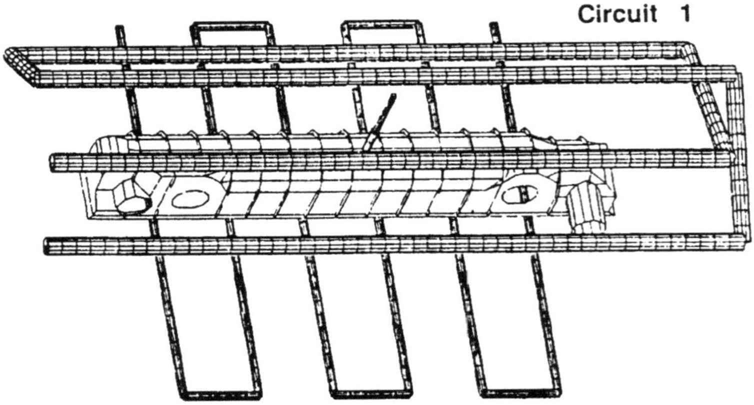

Fig. 7.1 Cooling lines related to the mold cavity (From Zheng et al. (1996), with permission from Sage Science Press)

$$\nabla^2 T = \frac{1}{\alpha_M}\frac{\partial T}{\partial t}, \tag{7.1}$$

where $\alpha_M = k_M/(\rho_M c_M)$ is the thermal diffusivity of the mold, k_M, ρ_M and c_M, are the thermal conductivity, density and specific heat of the mold, respectively.

The mold temperature is initially set to the coolant temperature, and subsequently the mold temperature field at the end of the previous cycle is used as the initial condition for the new cooling cycle.

The mold boundary Γ is comprised of the cavity surface Γ_p, the cooling channel surface Γ_c, and the external surface Γ_e. The boundary conditions are as follows:

Mold cavity surface

$$k_M\frac{\partial T}{\partial n} = -q, \quad \text{in}\,\Gamma_p, \tag{7.2}$$

Cooling channel surface

$$k_M\frac{\partial T}{\partial n} = -h_c(T - T_b), \quad \text{in}\,\Gamma_c, \tag{7.3}$$

Mold exterior surface

$$k_M\frac{\partial T}{\partial n} = -h_a(T - T_a), \quad \text{in}\,\Gamma_e. \tag{7.4}$$

In Eq. 7.2, q is the heat flux across the melt and mold interface, which is not known beforehand. The mold cooling analysis is essentially coupled with the transient heat transfer in the polymer melt during filling, packing, and cooling stages.

In Eq. 7.3, h_c represents the heat transfer coefficient between the mold and the coolant at the coolant's bulk temperature T_b. The heat transfer coefficient is given by

$$h_c = \frac{k_c Nu}{d_c}, \tag{7.5}$$

where k_c is the thermal conductivity of the coolant, d_c is the diameter of the cooling channel, and Nu is the Nusselt number. According to Dittus and Boelter (cf. Holman 1976):

$$Nu = 0.023\, Re^{0.8}\, \mathrm{Pr}^{0.4}, \tag{7.6}$$

where Re and Pr are the Reynolds number and the Prandtl number, respectively. Because both T and T_b in Eq. 7.3 can vary along the length of the cooling channel, an averaging process called the log mean temperature difference (LMTD) is adopted for use with Eq. 7.3. At each element of the cooling channel, let T_b^I and T_b^O stand for the inlet and outlet coolant temperatures, respectively, then we have

$$\Delta T = \frac{\Delta T_O - \Delta T_I}{\ln(\Delta T_O / \Delta T_I)}, \tag{7.7}$$

where $\Delta T_O = T - T_b^O, \Delta T_I = T - T_b^I$, and ΔT is the LMTD value for $T - T_b$.

In Eq. 7.4, T_a is the temperature of the ambient air and h_a represents the heat transfer coefficient between the mold and the ambient air. Approximately, $h_a \approx 10\,\mathrm{W/m^2K}$. It can also be evaluated from

$$h_a = \frac{k_a Nu}{L_m}, \tag{7.8}$$

where k_a is the thermal conductivity of the ambient air, L_m is the characteristic length of the mold exterior surface, which can be calculated from the area-to-perimeter ratio of the surface, and Nu is the Nusselt number which may be expressed by the following empirical equation for a variety of circumstances:

$$Nu = C(\mathrm{Gr\,Pr})^m, \tag{7.9}$$

where Gr is the Grashof number (the ratio of the buoyancy force to the viscous force) and Pr is the Prandtl number (the ratio of the kinematic viscosity to the thermal diffusivity) of air. The values of C and m depend on particular cases; for example, $C = 0.1$ and $m = 1/3$ for vertical planes, $C = 0.54$ and $m = 1/4$ for free convection from upper horizontal planes, and $C = 0.58$ and $m = 1/5$ for free convection from bottom horizontal planes (Holman 1976).

Because of the complexity in modeling the real mold exterior surfaces, one may approximate the mold exterior surfaces as an equivalent box or sphere.

Since heat lost through the mold exterior surface is very small in most injection molding applications, one can also treat the mold exterior surfaces as an infinite

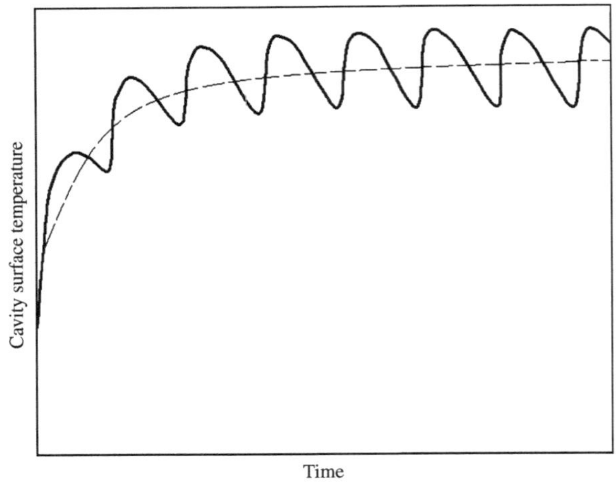

Fig. 7.2 The pattern of the transient mold temperature at cavity surface

adiabatic surface, as recommended by Rezayat and Burton (1990). This approximation does not require modeling of the mold exterior surface. The external boundary condition can be simply $T = T_\infty$, where T_∞ is an unknown constant that can be determined using an iterative algorithm based on the energy balance between the total heat gain from the polymer melt and the total heat loss through the cooling channels (Park and Kwon 1996).

7.3 Cycle-Average Simplification

The mold temperature in the continuous injection-molding operation can be separated into two components: the cycle-averaged component and the fluctuating component. After a short initial transient period, the cycle-averaged temperature reaches a steady state, and the fluctuating component of mold temperature is small compared to the cycle-average component (Fig. 7.2). Therefore, if one neglects the initial unsteady period and the fluctuating component, the cooling analysis can be done based on the steady-state cycle-averaged temperature only. This is an approximation but has been widely used (e.g., Himasekhar et al. 1992; Park and Kwon 1998; Fan et al. 2005; Zhou et al. 2009).

The steady-state cycle-averaged temperature is governed by the Laplace equation:

$$\nabla^2 T = 0, \tag{7.10}$$

The boundary condition (7.2) is modified to

$$k_M \frac{\partial T}{\partial n} = -\bar{q}, \quad \text{in } \Gamma_p, \tag{7.11}$$

where $\bar{q}$ is the cycle-averaged heat flux given by

$$\bar{q} = \frac{1}{t_T} \int_0^{t_T} q(t)dt. \tag{7.12}$$

Here t_T denotes the total cycle time defined as the sum of filling time, packing time, cooling time and mold opening time. The heat flux during the mold opening time is typically very small and may be neglected.

Chapter 8
Computational Techniques

8.1 Introduction

Analytical solutions to injection molding problems are very rare due to the complexities of the governing equations, the material behavior and the cavity geometry. To get useful results, we have to seek numerical solutions. In any numerical solution procedure, the governing equations are discretized to form a set of algebraic equations, possibly nonlinear, and computational algorithms are developed to solve the algebraic equations. Different discretization processes and different solution algorithms form a variety of numerical methods; each method has some advantage over the others in a certain class of problems. In this Chapter, we shall deal with several numerical methods including the finite element method, the finite difference method, the meshless particle method, and the boundary element method. However, the aim of this Chapter is not to provide in-depth discussions about the fundamental aspects of numerical methods or a comprehensive reference to the computer aided engineering software. Instead, the focus of the Chapter is to provide a guide to some special issues and computational techniques dealing with injection molding problems.

It is convenient to divide the whole procedure of injection molding simulation into three modules:

- Mold cooling analysis
- Filling/Packing/Residual-stress analysis (or simply called flow analysis)
- Structural analysis (for shrinkage and warpage)

The mold cooling analysis and the flow analysis are essentially coupled since the transient cavity wall temperature and heat flux are unknown in both analysis, although in practice one may only need a couple of iterative loops, depending on the required accuracy and efficiency.

This Chapter will discuss the flow analysis, structural analysis and the mold cooling analysis in sequence.

R. Zheng et al., *Injection Molding*, DOI: 10.1007/978-3-642-21263-5_8,

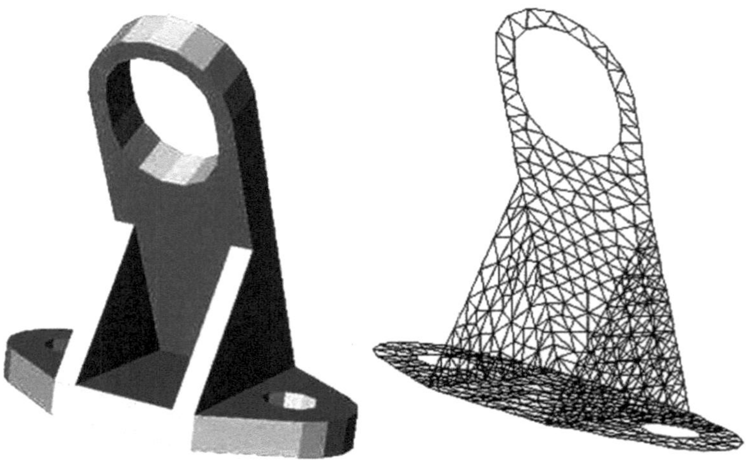

Fig. 8.1 An example of midplane mesh (Courtesy of Dr. Peter Kennedy)

8.2 Flow Analyses

8.2.1 Midplane Approach

8.2.1.1 Solving the Pressure Problem

The Hele-Shaw equation for the determination of pressure has been derived for a two-dimensional geometry. To solve the pressure problem for a thin cavity of general planar geometry in three-dimensional space, we use a finite-element (FE) representation on the midplane of the cavity (Fig. 8.1). Each element is assigned a thickness. The Hele-Shaw equation is discretized on each element using the local coordinate system associated with that element. The unknown node pressure and the volumetric flow rate are all scalar quantities and they are not linked to the coordinate system. In addition, we use a finite difference (FD) method to discretize the time- and gap-wise coordinates to solve the energy equation for the temperature field. In the following derivation of the FE/FD equations, only the cavity planar flow is considered. Derivation of the axisymmetric form of the equations for the runner flow can be done in the same manner. This approach deals with a 2-D pressure field, coupled to a 3-D temperature field, and therefore it is called a 2.5D simulation.

Let us assume a triangular element with nodes 1, 2 and 3 is located in the global coordinate system X_1, X_2 and X_3, as schematically shown in Fig. 8.2. We define two vectors $\boldsymbol{l}_3$ and $\boldsymbol{l}_2$ as

$$\boldsymbol{l}_3 = \left(X_1^{(2)} - X_1^{(1)}, X_2^{(2)} - X_2^{(1)}, X_3^{(2)} - X_3^{(1)}\right), \tag{8.1}$$

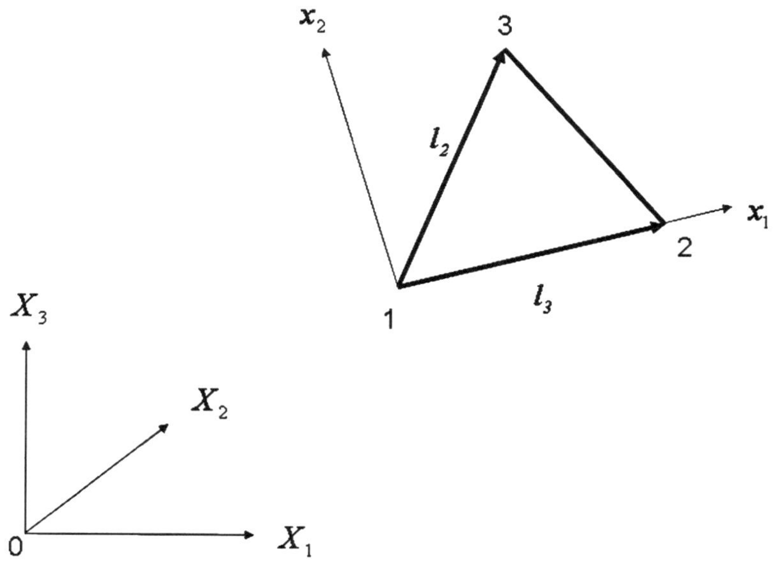

Fig. 8.2 Global (X_1, X_2 and X_3) and local (x_1, x_2 and x_3) coordinates of a triangular finite element

and

$$\boldsymbol{l}_2 = \left(X_1^{(3)} - X_1^{(1)}, X_2^{(3)} - X_2^{(1)}, X_3^{(3)} - X_3^{(1)} \right), \tag{8.2}$$

where the super scripts denote the nodal number (for example $X_1^{(3)}$ denotes the X_1 coordinate of node 3).

In Fig. 8.2, x_1, x_2 and x_3 represent the local coordinates, where node 1 is the origin. The local x_1-axis is along the direction of vector $\boldsymbol{l}_3$, i.e., the direction of node 1 pointing to node 2. The x_2-axis lies in the plane defined by nodes 1, 2 and 3, and points toward node 3 perpendicular to the x_1-axis. The x_3-axis is defined by $\boldsymbol{l}_3 \times \boldsymbol{l}_2$, which is normal to the element plane. The transformation from the global coordinates to the local element coordinates is as follows:

Node 1:

$$x_1^{(1)} = x_2^{(1)} = x_3^{(1)} = 0. \tag{8.3}$$

Node 2:

$$x_1^{(2)} = (\boldsymbol{l}_3 \cdot \boldsymbol{l}_3)^{1/2}, \tag{8.4}$$

$$x_2^{(2)} = x_3^{(2)} = 0. \tag{8.5}$$

Node 3:

$$x_1^{(3)} = \frac{\boldsymbol{l}_3 \cdot \boldsymbol{l}_2}{(\boldsymbol{l}_3 \cdot \boldsymbol{l}_3)^{1/2}}, \tag{8.6}$$

$$x_2^{(3)} = \left[\left(\boldsymbol{l}_2 - \frac{\boldsymbol{l}_3 \cdot \boldsymbol{l}_2}{\boldsymbol{l}_3 \cdot \boldsymbol{l}_3}\boldsymbol{l}_3\right) \cdot \left(\boldsymbol{l}_2 - \frac{\boldsymbol{l}_3 \cdot \boldsymbol{l}_2}{\boldsymbol{l}_3 \cdot \boldsymbol{l}_3}\boldsymbol{l}_3\right)\right]^{1/2}, \tag{8.7}$$

$$x_3^{(3)} = 0. \tag{8.8}$$

To describe the finite element formulation of the discretized Hele-Shaw equation, we write the Hele-Shaw equation in the following form

$$\nabla \cdot (S\nabla P) - K_P\frac{\partial P}{\partial t} - \Pi_T = 0, \quad \mathbf{x} \in \Omega. \tag{8.9}$$

We now discretize the domain Ω into a set of non-overlapping finite elements, for example, a collection of triangles denoted by Ω_e:

$$\Omega = \bigcup_e \Omega_e. \tag{8.10}$$

Application of the Galerkin weighted-residual method to Eq. 8.9 gives

$$\left\langle \nabla \cdot (S\nabla P), N_n^{(e)} \right\rangle - \left\langle \mathrm{K}_P\frac{\partial P}{\partial t}, N_n^{(e)} \right\rangle - \left\langle \Pi_T, N_n^{(e)} \right\rangle = 0, \tag{8.11}$$

where $\langle a, b \rangle$ denotes the inner product $\int_{\Omega_e} ab\, d\Omega_e$, $N_n^{(e)}$ is the shape function for node n in element e. We integrate the above equation by parts to obtain

$$\begin{aligned} &\int_{\Omega_e} S\nabla N_n^{(e)} \cdot \nabla P d\Omega_e - \int_{\partial\Omega_e} SN_n^{(e)}\nabla P \cdot nd(\partial\Omega_e) \\ &+ \int_{\Omega_e} \mathrm{K}_P N_n^{(e)}\frac{\partial P}{\partial t}d\Omega_e + \int_{\Omega_e} \Pi_T N_n^{(e)} d\Omega_e = 0, \end{aligned} \tag{8.12}$$

where $\partial\Omega_e$ is the boundary of Ω_e. The boundary integral gives the flow rate at node n of element e, i.e.,

$$\int_{\partial\Omega_e} N_n^{(e)}(S\nabla P \cdot \mathbf{n})d(\partial\Omega_e) = Q_n^{(e)}. \tag{8.13}$$

Next, on each of the triangular elements Ω_{e}, the pressure field is approximated by a simple piecewise varying function, using the shape function as the interpolation function.

$$P \approx \sum_{j=1}^{3} P_j^{(e)} N_j^{(e)}, \tag{8.14}$$

where the $P_j^{(e)}$ is the nodal value of pressure for node j of element e. Note that the three-node triangle is just one of the infinite series of triangular elements that can be specified. Higher-order triangular elements obtained by assigning additional nodes at the edges of the triangle or inside the triangle are also widely used, see Zienkiewicz et al. (2005) for details.

$$\sum_{j=1}^{3} P_j^{(e)} \int_{\Omega_e} S\nabla N_n^{(e)} \cdot \nabla N_j^{(e)} d\Omega_e + \sum_{j=1}^{3} \frac{\partial P_j^{(e)}}{\partial t} \int_{\Omega_e} \mathrm{K}_p N_n^{(e)} N_j^{(e)} d\Omega_e$$
$$= Q_n - \int_{\Omega_e} \Pi_T N_n^{(e)} d\Omega_e, \tag{8.15}$$

which can be rearranged in the form of simultaneous equations

$$\left[\mathbf{K}^{(e)}\right]\left\{\mathbf{P}^{(e)}\right\} + \left[\mathbf{C}^{(e)}\right]\left\{\dot{\mathbf{P}}^{(e)}\right\} = \left\{\mathbf{R}^{(e)}\right\}, \tag{8.16}$$

where $\left[\mathbf{K}^{(e)}\right]$ and $\left[\mathbf{C}^{(e)}\right]$ are element stiffness matrices resulting from the integration of $S\nabla N_n^{(e)} \cdot \nabla N_j^{(e)}$ and $\mathbf{K}_P N_n^{(e)} N_j^{(e)}$, respectively; $\left\{\mathbf{P}^{(e)}\right\}$ is the nodal pressure vector, and $\left\{\dot{\mathbf{P}}^{(e)}\right\}$ is its time derivative; $\left\{\mathbf{R}^{(e)}\right\}$is the right hand side vector. This system of equations is called the element equations.

The element equations are then assembled to form a global system of equations, considering the element interconnection requirement. Using a first order difference approximation to the time derivative of the pressure, the global system of equations can be written as

$$\left([\mathbf{K}] + \frac{1}{\Delta t}[\mathbf{C}]\right)\{\mathbf{P}(t+\Delta t)\} = \frac{1}{\Delta t}[\mathbf{C}]\{\mathbf{P}(t)\} + \{\mathbf{R}\}. \tag{8.17}$$

In the global system of equations, the total flow rates (contained in $\{\mathbf{R}\}$) are non-zero only at the injection nodes and at the nodes on the flow front while the flow front is progressing. Total flow rates are zero at all the filled nodes, because the flow into a filled node must be equal to the flow out from the node. Having established the global system of equations, the relevant boundary conditions can then be applied. In the filling stage, the flow rate is specified at the injection notes and a zero pressure is imposed at nodes on the flow front. In the packing stage, the pressure as a function of time is specified at the injection nodes. One can also specify a switch-over criterion by percent volume to control the transition from the flow-rate boundary condition to the pressure boundary condition. For example, if the value specified is 100%, the flow-rate boundary condition is used in the entire filling stage; if the value specified is less than 100%, then the flow-rate boundary condition is switched to a pressure condition early in the filling stage.

For non-Newtonian flows, the stiffness matrix depends on shear rate and temperature and thus on pressure. The algebraic system is non-linear and it should be solved iteratively.

There are some constraints on the shape functions in order to have a consistent finite element formulation. Equation 8.12 demands that the shape function $N_n^{(e)}$ be at least linear in the spatial coordinates. The simplest and most widely used shape function in our case is perhaps the piecewise linear interpolation function as described in Kennedy (1995). With this shape function, one can achieve C^0 continuity in pressure, that is, the pressure field variable is continuous at element interface, but its gradients are not. The pressure gradient field is piecewise constant over the elements and is discontinuous across element interfaces. Consequently, the resulting velocity and shear rate fields are not continuous across element boundaries. An example of triangular elements with higher order shape functions can be found in Hieber and Shen (1980).

8.2.1.2 Solving the Temperature Problem

The temperature profile across the thickness as a function of time is determined by solving the energy equation. A finite difference (FD) method is employed, where the differential grid points are located at each vertex node of the triangular elements, across the full-gap thickness. In the FD representation of the energy equation, an implicit form is used for the gap-wise thermal conduction term. The convective, viscous dissipation and heat source terms are evaluated at the previous time t. Thus, the temperature can be determined at the new time step $t + \Delta t$.

If the pressure field is C^0 continuous, the convective, viscous dissipation and heat source terms are not continuous at a given node due to the discontinuity in the pressure gradients across the element. These terms are therefore evaluated at the centroid of each element, and then a weighted averaging process is used to determine the nodal values based upon the contributions from adjacent elements. However, for numerical stability, in the calculation of the nodal value of the convective term, only contributions from the adjacent upstream elements are considered. This is known as the upwinding scheme.

Suppose that the gap-wise temperature field is covered with a non-uniform grid of size $\Delta z_j = z_j - z_{j-1}$ (where the local z-direction is the same as the local x_3-direction), together with a time increment $\Delta t = t_{n+1} - t_n$. We define

$$T_j^n = T(z_j, t_n), \tag{8.18}$$

and

$$r_j = \frac{\Delta z_{j+1}}{\Delta z_j}, \tag{8.19}$$

then we have

$$\frac{\partial T}{\partial t} \approx \frac{T_j^{n+1} - T_j^n}{\Delta t}, \tag{8.20}$$

$$\frac{\partial^2 T}{\partial z^2} \approx \frac{2}{1+r_j}\left[\frac{\frac{1}{r_j}T_{j+1}^{n+1} - \left(1+\frac{1}{r_j}\right)T_j^{n+1} + T_{j-1}^{n+1}}{(\Delta z_j)^2}\right]. \tag{8.21}$$

Assuming that the thermal conductivity is independent of z, we have the following FD discretized energy equations:

$$-\frac{M}{r_j}T_{j+1}^{n+1} + \left[1 + M\left(1+\frac{1}{r_j}\right)\right]T_j^{n+1} - MT_{j-1}^{n+1} = T_j^n + R_T\Delta t, \tag{8.22}$$

where

$$M = \frac{2k\Delta t}{(1+r_j)\rho c_p(\Delta z_j)^2}, \tag{8.23}$$

$$R_T = -(\mathbf{u}\cdot\nabla T)_j^n + \frac{1}{\rho c_p}(\tau : \mathbf{D})_j^n + \dot{Q}_j^n. \tag{8.24}$$

At each time step, the temperature problem and the pressure problem are solved independently, and the coupling is enforced by iteration. The procedure is as follows:

1. Solve pressure and flow fields. The stiffness matrices of the system of equations are calculated based on the pressure and temperature data from the previous time step.
2. After the pressure has been calculated, the new values of velocity, shear rate and viscosity are evaluated and used to calculate the convective and viscous dissipation terms. Calculations of Eq. 8.22 are performed with the finite difference method and give the temperature field. With the new temperature field, an updated viscosity field is calculated. Steps 1 and 2 are repeated until convergence is achieved.
3. After the pressure and temperature calculations have converged, the velocity and flow rate are calculated. The flow front is advanced accordingly, if the cavity is not fully filled. The numerical methods for advancing the flow front will be discussed in the next section.

8.2.2 Advancement of the Flow Front

An important task in the analysis of filling stage is to determine the location of the moving flow front (also called the melt front). The calculations can be based on either a Lagrangian method or an Eulerian method. In the Lagrangian method, only the fluid domain is meshed and the location of the fluid front is represented by the frontal boundary of the moving mesh. The family of Lagrangian methods includes the so-called arbitrary Lagrangian Eulerian (ALE) method (Hughes et al. 1981)

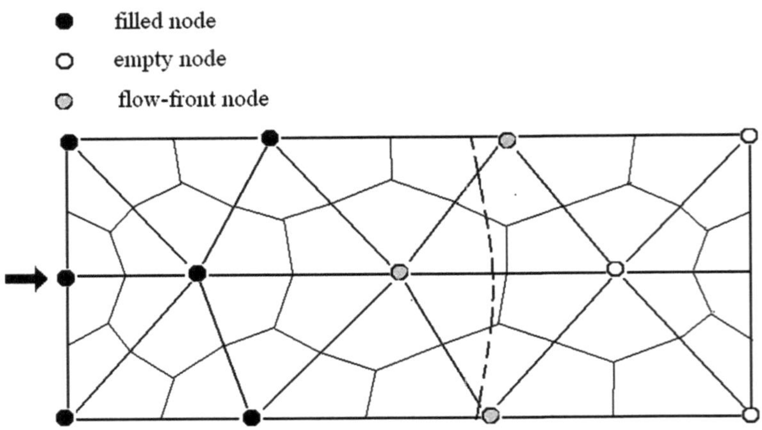

Fig. 8.3 Control-volume and node definition for advancement of the flow front

where only the boundary of the mesh is moving with the fluid, while the internal nodes shift their positions to maintain reasonably shaped meshes. On the other hand, in the Eulerian method, the whole cavity is meshed, and the mesh is fixed. The free surface (or interface) is considered as a scalar field to be solved.

The Lagrangian method is limited by the mesh distortion and the need for a time-consuming remeshing, and hence the Eulerian method is favored for our purpose. We will describe below three different approaches based on the Eulerian method.

8.2.2.1 Control-Volume Method

Following the idea of flow analysis network (FAN), introduced by Tadmor et al. (1974), Chiang et al. (1991) described the following scheme using control volumes. In this method, one divides each triangular element into three sub-areas by linking the centroid of the element to the midpoints of its three edges. For each vertex node, the sum of all subareas containing that node defines a polygonal control volume. For each control volume, a parameter f_{CV} is introduced, defined as the ratio of the melt-occupied volume to the total control volume. The vertex nodes can be classified into three categories (Fig. 8.3), including

(i) *filled nodes*, for which the associated control volume is fully filled with melt ($f_{CV} = 1$);
(ii) *empty nodes*, for which the associated control volume is totally empty ($f_{CV} = 0$);
(iii) *flow-front nodes*, for which the associated control volume is partially filled with melt ($0 < f_{CV} < 1$).

The entrance node is always treated as a filled node. After having obtained the nodal pressures, one calculates the net flow rates into each partially filled control volume, and updates the corresponding value of f_{CV} for a given elapsed time interval. One can choose the time step such that within the time interval only one flow-front node gets filled, with all the unfilled nodes connected to the newly filled node becoming new flow-front nodes.

The scheme has two major drawbacks. One is the dependence on the number and shape of the elements. The second is oscillation of the melt front.

8.2.2.2 Volume of Fluid (VOF) Method

The Volume of Fluid (VOF) method, as introduced by Hirt and Nichols (1981), is based on the mass conservation principle. Similar to the control-volume method, in the VOF method, the whole domain can also be divided into control volumes, each of which is associated to an element node, and the volume fraction of fluid in each control volume is defined. The flow front is advanced by solving the following transport equation:

$$\frac{\partial F}{\partial t} + \mathbf{u} \cdot \nabla F = 0, \tag{8.25}$$

where F is a concentration function taking the value of the volume fraction of fluid in a control volume. F is unity in the fully filled control volume and zero in the empty control volume. Control volumes with values of F between 0 and 1 must then contain the flow front surface. In the equation, $\mathbf{u}$ is the fluid velocity vector. As the Hele-Shaw approximation treats the melt front as being flat in the gap-wise direction, we can replace $\mathbf{u}$ by the gap-wise averaging velocity $\bar{u}$. Equation 8.25 is solved simultaneously with the equations of motion, and F moves with the fluid.

In this technique, the calculation domain is treated as a two-phase system. One phase consists of the liquid filled region, and the other consists of a void region. The flow front is regarded as the interface separating the two phases. The void is assumed to contain a virtual fluid, such as gas, with a set of virtual physical properties. Physical parameters in the governing flow equations are expressed using the mixture law weighed by F. for example, for the density, ρ, one has

$$\rho = \rho_f F + \rho_g (1 - F), \tag{8.26}$$

where ρ_f and ρ_g are the values of the density in the filled liquid and the gas, respectively. Using such an expression in the flow equations, one can solve the flow problem for the entire domain. The solution scheme starts from a known distribution of the function F. At the end of each time step new values of F are obtained and the position of the flow front is updated.

Kietzmann et al. (1998) discussed discretization schemes for the solution of Eq. 8.25. The equation is a typical convection equation and an upwinding is required for numerical stability. However, the upwind discretization may result in numerical diffusion that blurs the flow front interface. A hybrid implicit scheme that combines upwind differencing and central differencing is therefore suggested.

8.2.2.3 Level Set Method

The level set method was proposed by Osher and Sethian (1988), and it has been used in a variety of interface evolution problems, such as multiphase flow (e.g., Zhao et al. 1996), free surface flow of liquid film (e.g., Dou et al. 2004), and moving flow front in a filling process (e.g., Dou et al. 2007). Books describing the level set method are available (e.g., Sethian 1999; Osher and Fedkiw 2002).

Suppose the computational domain is Ω. We divide it into two subdomains: the interior domain Ω_{interior} and the external domain Ω_{exterior}. The interior domain Ω_{interior} represents the liquid filled region, while the external domain Ω_{exterior} is the empty or gas filled region. The interface between the interior and exterior regions is denoted by $\partial\Omega$.

A distance function $d(\mathbf{X})$, $\mathbf{x} \in \Omega$, is defined as

$$d(\mathbf{x}) = \min(|\mathbf{x} - \mathbf{x}_l|) \quad \text{for all} \quad \mathbf{x}_l \in \partial\Omega, \tag{8.27}$$

implying that $d(\mathbf{x}) = 0$ on the interface.

The basic idea of the level set method is to define a *signed distance function* $\Phi(\mathbf{x}, t)$ in the computational domain, from which the zero level isocontour of the function is the interface representing the position of the flow front. Initially, Φ equals the distance from any point $\mathbf{x}$ to the initial $\partial\Omega$, negative inside filled region and positive outside the filled region, where the initial $\partial\Omega$ represents the inlet boundary. That is, for all $x_l \in \partial\Omega$,

$$\Phi(\mathbf{x}, t = 0) = \begin{cases} -\min(|\mathbf{x} - \mathbf{x}_l|) & \text{for } \mathbf{x} \in \Omega_{\text{interior}} & \text{(the liquid filled regions)}, \\ 0 & \text{for } \mathbf{x} \in \partial\Omega & \text{(the liquid-gas interface)}, \\ \min(|\mathbf{x} - \mathbf{x}_l|) & \text{for } \mathbf{x} \in \Omega_{\text{exterior}} & \text{(the gas filled regions)}, \end{cases} \tag{8.28}$$

which is continuous over the whole computational domain. In addition, the signed distance function has the following property:

$$|\nabla\Phi| = 1. \tag{8.29}$$

On the level set, the interface at any instant corresponds to the contour $\Phi(\mathbf{x}, \text{t}) = 0$. Since the interface moves with the fluid particles, the evolution of Φ is then given by the following convection equation:

$$\frac{\partial\Phi}{\partial t} + \mathbf{u} \cdot \nabla\Phi = 0, \tag{8.30}$$

where $\mathbf{u}$ is the fluid velocity.

The normal unit vector on the interface, drawn from the liquid to the gas, can be expressed as

$$\mathbf{n} = \left.\frac{\nabla\Phi}{|\nabla\Phi|}\right|_{\Phi=0}. \tag{8.31}$$

Thus Eq. 8.30 can also take the form

$$\frac{1}{|\nabla\Phi|}\frac{\partial\Phi}{\partial t}+\mathbf{u}\cdot\mathbf{n}=0. \tag{8.32}$$

where $\mathbf{u}\cdot\mathbf{n}$ is the normal velocity of fluid particles on the interface. Equations 8.30 or 8.32 can be solved by a time marching scheme. After each time step, the zero level set function should represent the new position of the interface. Since the equation is of the hyperbolic type, stabilization techniques such as upwinding are usually employed.

Because of the numerical approximation, the level set function $\Phi(\mathbf{x}, t)$ may not remain a signed distance function at later time steps, in particular after a long simulation time. Therefore, after the level set function has been advected, it must be re-initialized so that it remains a distance function. The re-initialization can be achieved by solving the following partial differential equation (Sussman et al. 1994, 1998; Sethian and Smereka 2003):

$$\frac{\partial\Phi}{\partial\tau}=sign(\Phi)(1-|\nabla\Phi|), \tag{8.33}$$

with initial conditions

$$\Phi(\mathbf{x},0)=\Phi_0(\mathbf{x}), \tag{8.34}$$

where sign is the sign function, τ is the artificial time, and $\Phi_0(\mathbf{x})$ is the initial value of Φ ($\boldsymbol{x}$, t) given at the beginning of calculation for the entire domain. Solving the equation to steady state with the artificial time τ provides a new value for Φ that satisfies $|\nabla\Phi|=1$, since the steady state is reached when the right-hand side approaches zero. The procedure is to stop the level set calculation periodically and solve Eq. 8.33 until reaching a steady state. Dou et al. (2007) suggest doing re-initialization after every time step.

During the solution of the flow governing equations in the whole computational domain Ω, the abrupt change of fluid properties such as density and viscosity across the liquid-gas interface may cause numerical instability. A finite thickness of the interface is hence defined, which is bound within $[-\varepsilon, \varepsilon]$, and a property averaging is used to smooth the fluid properties. Assume that the liquid and the gas properties are denoted by M_{liquid} and M_{air} respectively (where M can be, for example, density or viscosity). The averaged material properties depending on Φ are expressed as

$$M(\Phi)=M_{\text{liquid}}+(M_{\text{gas}}-M_{\text{liquid}})H_\varepsilon(\Phi), \tag{8.35}$$

with

$$H_\varepsilon(\Phi)=\begin{cases}0 & \text{if} \quad \Phi<-\varepsilon\\ \frac{1}{2}\left[1+\frac{\Phi}{\varepsilon}+\frac{1}{\pi}\sin(\pi\Phi/\varepsilon)\right] & \text{if} \quad |\Phi|\le\varepsilon\\ 1 & \text{if} \quad \Phi>\varepsilon\end{cases}, \tag{8.36}$$

which is known as the smeared-out Heaviside function (Osher and Fedkiw 2002). The use of the function allows for a smooth transition of the fluid properties from the liquid to the gas or void.

8.2.2.4 Weld Lines

When two flow fronts collide, a weld line is formed. Weld lines are often caused by multiple gates, inserts and thickness variations. Such weld lines not only harm the aesthetics of the product but also weaken the local mechanical strength. Hagerman (1973) attributes the weakness of weld-line tensile strength to three primary factors: poor bonding adhesion, frozen-in parallel molecular orientation, and existence of V-notches. Although the weld lines are not always avoidable, their locations can be predicted by a flow analysis. Thus one can redesign the mold to position weld lines in noncritical areas, considering both structural and aesthetic demands (see, for example, Zhai et al. 2006).

Zhou and Li (2004) proposed a weld-line detector algorithm applied to triangular finite-element mesh. The algorithm consists of the following steps:

1. The first step is the computation of the time of flow fronts reaching each finite-element node (named as the "reaching time" of the node) based on flow-front advancing simulation results.
2. The second step is the detection of the so-called initial meeting node defined as the first contact point of two flow fronts meeting. For a finite-element node of the mesh, if there are four surrounding triangles at each of which the reaching time of the node is shorter than that of one of the other two nodes but longer than that of the other one on the same triangle, the node is an initial meeting node. The idea is illustrated in Fig. 8.4, where curves F_1 and F_2 represent two flow fronts. Node N_1 is an initial meeting node, and the four surrounding triangles satisfying the above mentioned condition are $\Delta N_1N_2N_3$, $\Delta N_1N_4N_5$, $\Delta N_1N_6N_7$ and $\Delta N_1N_7N_8$, in which we have $t_{N_2} < t_{N_1} < t_{N_3}$, $t_{N_5} < t_{N_1} < t_{N_4}$, $t_{N_6} < t_{N_1} < t_{N_7}$ and $t_{N_8} < t_{N_1} < t_{N_7}$, respectively, where t_{N_k} is the reaching time of node N_k.
3. The last step is to expand from the initial meeting node to obtain the entire weld lines.

Weld-line prediction is sensitive to the mesh density. When information of weld-line locations is required, it is essential to use a sufficient fine mesh in the areas where weld-lines are likely to occur.

8.2.3 *Fountain Flow Effect*

Fountain flow is a phenomenon encountered in the advancing front of injection molding flow. Figure 8.5 shows a cross section of the flow channel near the flow front. One observes the flow with the frame of reference travelling with the

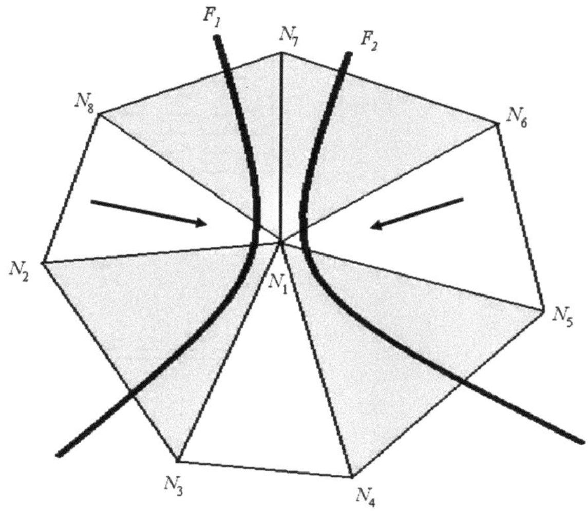

Fig. 8.4 An illustration of the method of Zhou and Li (2004) for detection of the initial meeting node. The *arrows* represent the advancing directions of the two meeting flow fronts

velocity of the advancing meniscus, so that the flow front is stationary while the cavity wall moves with a negative velocity. The velocity field shown in the figure was obtained from a boundary element simulation (Zheng 1991). One can see that in the central region the fluid particles decelerate as they approach the front and spill over toward the walls, exhibiting a rolling-type motion. The region of parallel flow located upstream of the free surface is divided in a core subregion and a skin subregion. In the core subregion the relative velocity is positive, and in the skin subregion the relative velocity is negative. Between the two subregions is a neutral line where the relative velocity is zero.

Previous numerical studies have been performed by several authors (Kamal et al. 1986, 1988; Mavridis et al. 1986, 1988; Coyle et al. 1987; Zheng et al. 1990; Jin 1993; Sato and Richardson 1995; Bogaerds et al. 2004; Baltussen et al. 2010; Mitsoulis 2010). These investigations have shown that the fountain flow has significant effects on the melt-front temperature, and also on the molecular or fiber orientation distributions in the skin region near the cavity wall.

Unfortunately, the Hele-Shaw approximation cannot represent the curvature of the flow front in the thickness direction. The flow front has to be treated as being flat in the thickness direction. One has to seek for approximate methods able to capture the fountain flow effect without resolving all the flow details. A crude way of dealing with the fountain effect on the melt-front temperature is to assume it is uniform and set it equal to the temperature in the core region behind the advancing front during the filling stage. This approach, however, presents mathematical inconsistency and lacks energy balance in the filling process, as pointed out by Crochet et al. (1994). Dupret and Vanderschuren (1988) have developed a superior model by flipping over the material points to mimic the basic kinematics of the fountain flow. As shown in Fig. 8.6, where the segment AB is the front zone,

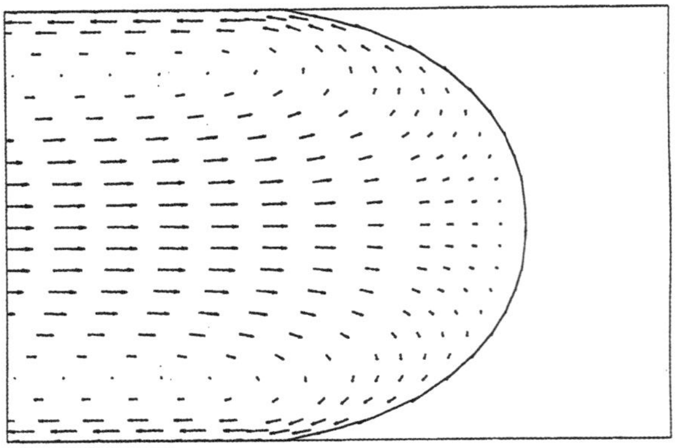

Fig. 8.5 The fountain flow velocity vector field in the moving frame of reference (Reproduced from Zheng (1991))

x-axis represent the symmetry plane, and point C indicated the location (of the neutral line. Heat diffusion and viscous heating are neglected in the front zone. A material point from the core subregion entering the front at level ζ_1 reappears at level ζ_2 in the skin subregion, while keeping the same temperature, that is,

$$T(\zeta_2) = T(\zeta_1), \tag{8.37}$$

The levels ζ_1 and ζ_2 are related by the following condition, based on mass conservation,

$$\int_0^{\zeta_1} (\mathbf{u} - \bar{\mathbf{u}}) \cdot \mathbf{n} d\zeta = \int_{\zeta_2}^{h/2} (\mathbf{u} - \bar{\mathbf{u}}) \cdot \mathbf{n} d\zeta, \tag{8.38}$$

where $\mathbf{n}$ is the outward normal of the front surface, $\mathbf{u}$ is the velocity and $\bar{\mathbf{u}}$ is the average velocity given by

$$\bar{\mathbf{u}} = \frac{1}{h} \int_{-h/2}^{h/2} \mathbf{u} dx_3. \tag{8.39}$$

The same method can also be used for the simulation of fountain effect on orientation (Crochet et al. 1994).

8.2.4 Dual Domain Approach

The midplane FE approach is by far the most efficient numerical method for the simulation of injection molding of thin-walled parts. However, the preparation of a midplane mesh can take a considerable amount of time.

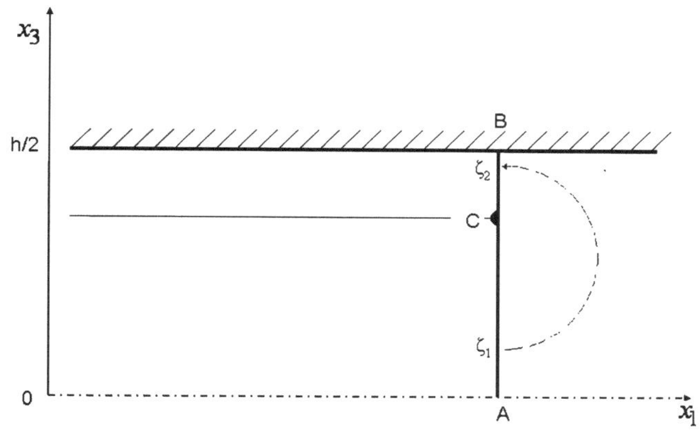

Fig. 8.6 Simplified fountain flow model (Reproduced from Dupret and Vanderschuren (1988), with permission from John Wiley and Sons)

The recent development of computer aided design (CAD) technology has led to the increasing use of surface and solid modeling that allows the creation of photorealist models, in which the outer component geometry is fully described in a 3-D space. Some efforts have been made to develop an automatic midplane generator that can deduce the midplane mesh from the CAD created 3-D geometry (see, for example, Kennedy and Yu 1997). This technique, however, has some difficulties with small features of the structure such as bosses. The generated mesh often has discontinuities that need to be fixed manually.

The limitations of the automatic midplane generation stimulated the innovation of the so-called dual domain approach (Yu and Thomas 2000; Yu et al. 2004). This approach uses the external mesh on a 3-D geometry. It still solves the Hele-Shaw equation but eliminates the need for a midplane mesh.

Figure 8.7 illustrates the basic idea. Figure 8.7a shows a rectangular plate injected at its center. Figure 8.7b shows the flow in a cross-cross section. Figure 8.7c is the midplane representation of the flow, and Fig. 8.7d is the dual domain representation of the flow, where two analyses are performed simultaneously on the top and bottom surface meshes.

Some more complex cases exist. An example highlighting the complexity is illustrated in Fig. 8.8, which shows the cross-section of a ribbed plate. Owing to the use of the surface mesh, when the flow hits the rib, at the bottom surface it continues flowing to the right, and at the top surface it goes up the rib and will flow over the top surface and then goes down along the opposite surface of the rib. Such an analysis would be physically incorrect. The solution is to impose synchronization by matching opposite elements and assigning the same pressure to the modes A, B, C and D. Flow now emanates from D as shown in Fig. 8.9, so the flow goes up the rib on both sides as required.

The dual-domain technique is still classified as a 2.5D simulation.

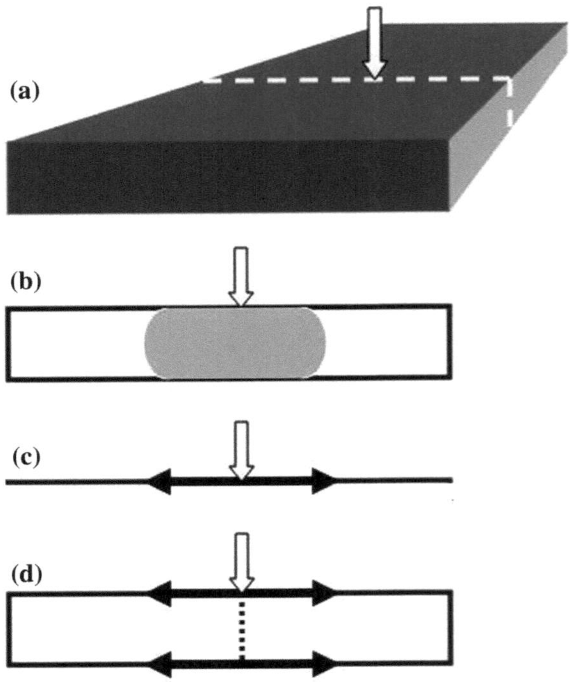

Fig. 8.7 Sketch of dual domain flow analysis: **a** injection into a rectangular plate; **b** flow of molten polymer in a cross-section of the cavity; **c** midplane representation; **d** dual domain concept (Courtesy of Dr. Peter Kennedy)

8.2.5 Three-Dimensional Finite Element Method

The earliest attempt of thee-dimensional (3-D) analysis of the filling stage in injection molding was made by Hétu et al. (1995, 1998) using a finite element method. The 3-D simulation does not make use of the Hele-Shaw approximation. Instead, it solves the conservation equations and constitutive equations over a three-dimensional solution domain. For generalized Newtonian fluids, the primary unknowns to be solved are velocity, pressure and temperature fields. For visco-elastic fluids, the stress components may be additional primary unknowns. The equations are nonlinear, and there is a coupling between thermal and mechanical equations. To solve simultaneously for the whole set of primary variables (usually by means of a Newton-Raphson iteration scheme) usually requires huge computer memory. A simple approach for this type of problem is an alternating solution scheme. In this scheme, one obtains the initial flow kinematics by solving an isothermal linear flow problem. With the known kinematics, one solves the energy equation for the temperature field (the constitutive equation can also be solved for stresses if necessary). The kinematics are then updated with the new temperature (and the stress) field, and the iterative procedure continues until convergence is achieved. The alternating solution scheme is allied to the Picard (fixed-point) iterative method.

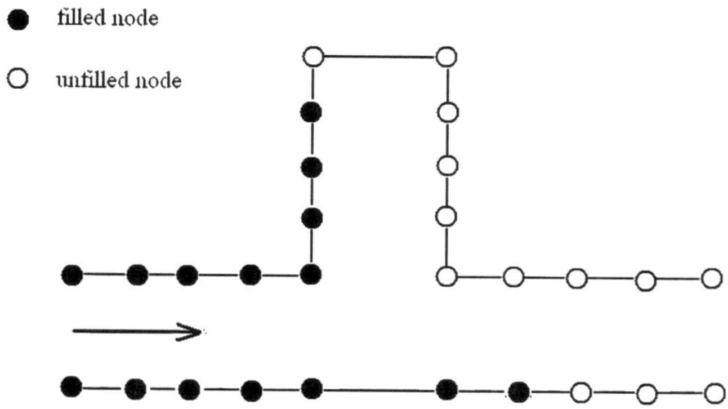

Fig. 8.8 Sketch of problem in dual domain flow analysis for a ribbed plate

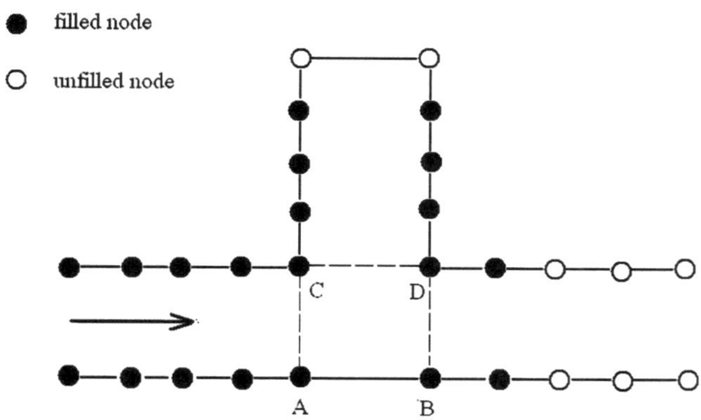

Fig. 8.9 Assignment of identical pressures at nodes to synchronize flow fronts at a rib

We now begin with discretization of the flow problem. For the sake of simplicity, we neglect inertia and gravity, and assume that the fluid is incompressible during the mold filling. Further, we choose the generalized Newtonian fluid to be the constitutive model. With these simplifications, the corresponding governing equations for the flow in the computational domain Ω are:

$$\nabla \cdot \boldsymbol{\sigma} = 0\,, \quad \text{in}\,\Omega, \tag{8.40}$$

$$\nabla \cdot \mathbf{u} = 0\,, \quad \text{in}\,\Omega, \tag{8.41}$$

where $\mathbf{u}$ is the velocity vector and $\boldsymbol{\sigma}$ is the total stress tensor, which is given by

$$\boldsymbol{\sigma} = -P\mathbf{I} + \eta\left(\nabla\mathbf{u} + \nabla\mathbf{u}^T\right). \tag{8.42}$$

Here, P is the pressure, $\mathbf{I}$ is the unit tensor (δ_{ij}), and η is the shear-rate and temperature dependent viscosity.

On the boundary $\partial\Omega$ of the domain Ω, the boundary conditions can be either velocity or traction boundary conditions.

$$\mathbf{u} = \mathbf{u}_0, \quad \text{on } \partial\Omega_u, \tag{8.43}$$

$$\mathbf{t} = \boldsymbol{\sigma} \cdot \mathbf{n} = \mathbf{t}_0, \quad \text{on } \partial\Omega_\sigma, \tag{8.44}$$

where $\mathbf{t}$ is the traction vector (also called the total normal stress), $\mathbf{n}$ is the outward normal, $\mathbf{u}_0$ and $\mathbf{t}_0$ are the prescribed velocity and traction, respectively. In the mold filling modeling, a zero velocity boundary condition is assumed at the mold/polymer interface and a zero traction boundary condition is assumed at the flow-front boundary. At the inlet boundary, either the injection velocity or the traction conditions are imposed. Note that the boundary condition in terms of the traction is different from the pressure boundary conditions. In many practical cases, in the inlet boundary, we require that the tangential component of traction vanishes there and only have non-zero normal traction components specified. In this case, from the constitutive Eq. 8.42, one has

$$\mathrm{t}_n = -P + 2\eta \frac{\partial u_n}{\partial n}, \tag{8.45}$$

where u_n is the component of velocity normal to the boundary, and $\partial/\partial n$ stands for the outward normal gradient at the boundary. If $\eta\partial u_n/\partial n$ is negligibly small, then the traction boundary condition is essentially the same as the pressure boundary condition.

The finite element procedure begins with the division of the computational domain Ω into a set of non-overlapping elements, for example, a collection of tetrahedral elements Ω_e. Since the Eulerian method is used, the elements are fixed in space. Methods such as the VOF or the level set method is used to determine the moving flow front.

Let V and Q denote the spaces of admissible functions for velocity and pressure. For test functions $\mathbf{w} \in \mathrm{V}$ and $q \in \mathrm{Q}$, we try to satisfy

$$\langle \nabla \cdot \boldsymbol{\sigma}, \mathbf{w} \rangle_\Omega = \mathbf{0}, \tag{8.46}$$

$$\langle \nabla \cdot \mathbf{u}, q \rangle_\Omega = 0, \tag{8.47}$$

where $\langle , \rangle$denotes the inner product. We shall use the following notation:

$$\langle s_1, s_2 \rangle_\Omega = \int_\Omega s_1 s_2 d\Omega, \tag{8.48}$$

$$\langle \mathbf{v}_1, \mathbf{v}_2 \rangle_\Omega = \int_\Omega \mathbf{v}_1 \cdot \mathbf{v}_2 d\Omega, \tag{8.49}$$

$$\langle \mathbf{v}_1, \mathbf{v}_2 \rangle_{\partial\Omega} = \int_{\partial\Omega} \mathbf{v}_1 \cdot \mathbf{v}_2 d\partial\Omega, \tag{8.50}$$

$$\langle \mathbf{T}_1, \mathbf{T}_2 \rangle_{\Omega} = \int_{\Omega} \mathbf{T}_1 : \mathbf{T}_2 d\Omega, \tag{8.51}$$

where s_1 and s_2 are scalar functions, $\mathbf{v}_1$ and $\mathbf{v}_2$ are vector functions, and $\mathbf{T}_1$ and $\mathbf{T}_2$ are tensor functions.

We integrate Eq. 8.46 by part, with help of the Gauss-Green theorem, to obtain

$$\langle \boldsymbol{\sigma}, \nabla \mathbf{w} \rangle_{\Omega} - \langle \boldsymbol{\sigma} \cdot \mathbf{n}, \nabla \mathbf{w} \rangle_{\partial\Omega} = 0, \tag{8.52}$$

which is rewritten as, according to Eq. 8.42

$$-\langle P, \nabla \cdot \mathbf{w} \rangle_{\Omega} + \langle 2\eta(\nabla \mathbf{u} + \nabla \mathbf{u}^T), \nabla \mathbf{w} \rangle_{\Omega} = \langle \sigma \cdot \mathbf{n}, \nabla \mathbf{w} \rangle_{\partial\Omega}. \tag{8.53}$$

This is known as the *weak formulation*. Since the boundary condition for the traction ($\mathbf{t} = \boldsymbol{\sigma} \cdot \mathbf{n}$) is directly incorporated in the weak formulation, the traction boundary condition is sometimes called the *natural boundary condition*.

On each tetrahedral element Ω_e, the velocity and pressure are approximated by simple piecewise varying functions. We choose the MINI-element, which has been applied to 3D injection molding simulations by Pichelin and Coupez (1998), and Yu and Kennedy (2004), following the pioneering work of Arnold et al. (1984). The velocity and the pressure filed are discretized, respectively, by

$$\mathbf{u}(\mathbf{x})|_{\Omega_e} = \sum_{k=1}^{4} N_k(\mathbf{x})\mathbf{u}_k^{(e)} + \Psi_b(\mathbf{x})\mathbf{u}_b, \tag{8.54}$$

$$P(\mathbf{x})|_{\Omega_e} \approx \sum_{k=1}^{4} N_k(\mathbf{x})P_k^{(e)}, \tag{8.55}$$

where $\mathbf{u}_k^{(e)}$ and $P_k^{(e)}$ are the nodal unknowns as the vertices of the tetrahedral element, $\mathbf{u}_b^{(e)}$ is the so-called "bubble" velocity at the internal node (Fig. 8.10), N_k is the standard linear shape function, and Ψ_b is a non-linear *bubble function* which is equal to 1 at the center of the tetrahedral element and equal to 0 at the boundary of the element. The approximation for the velocity field is said to be *linear enriched*, described by P^{1+}–C^0, and the approximation for the pressure is linear, of the P^1–C^0. Here P_k is referring the order of the polynomials of the shape function N_n, and C^k, as mentioned before, is referring the continuity properties of the approximation. The MINI element satisfies the Babuska-Brezzi stability condition (Arnold et al. 1984).

By substituting the above approximations into Eqs. 8.47 and 8.53, we obtain a system of equations in the following matrix form:

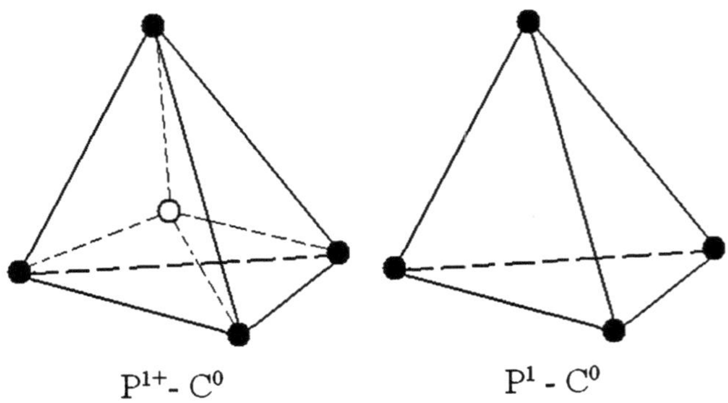

Fig. 8.10 MINI-element, with linear interpolation and bubble enrichment for velocity, and with linear interpolation for pressure

$$\left(\begin{bmatrix} \mathbf{K}_{11} & \mathbf{0} & \mathbf{K}_{13} \\ \mathbf{0} & \mathbf{0} & \mathbf{0} \\ \mathbf{K}_{31} & \mathbf{0} & \mathbf{0} \end{bmatrix} + \begin{bmatrix} \mathbf{0} & \mathbf{0} & \mathbf{0} \\ \mathbf{0} & \mathbf{K}_{22}^{(b)} & \mathbf{K}_{23}^{(b)} \\ \mathbf{0} & \mathbf{K}_{32}^{(b)} & \mathbf{0} \end{bmatrix}\right) \begin{Bmatrix} \mathbf{V}^{(l)} \\ \mathbf{V}^{(b)} \\ \mathbf{P} \end{Bmatrix} = \begin{Bmatrix} \mathbf{R} \\ \mathbf{R}^{(b)} \\ \mathbf{0} \end{Bmatrix}, \tag{8.56}$$

where the stiffness matrix includes two parts, the first part arises from the linear interpolation, and the second from the bubble enrichment; $\{\mathbf{V}^{(l)}, \mathbf{V}^{(b)}, \mathbf{P}\}^{\mathrm{T}}$ is the unknown solution vector for the linear part of velocity, bubble enriched velocity and pressure, and $\{\mathbf{R}, \mathbf{R}^{(b)}, \mathbf{0}\}^{\mathrm{T}}$ is the usual right-hand side vector. The bubble enriched velocity vector can be written in the condensation form:

$$\mathbf{V}^{(b)} = -\left[\mathbf{K}_{22}^{(b)}\right]^{-1}\left(\mathbf{K}_{23}^{(b)}\mathbf{P} - \mathbf{R}^{(b)}\right). \tag{8.57}$$

Thus, the system of Eq. 8.56 reduces to

$$\begin{bmatrix} \mathbf{K}_{11} & \mathbf{K}_{13} \\ \mathbf{K}_{31} & -\mathbf{K}_{32}^{(b)}(\mathbf{K}_{22}^{(b)})^{-1}\mathbf{K}_{23}^{(b)} \end{bmatrix} \begin{Bmatrix} \mathbf{V}^{(l)} \\ \mathbf{P} \end{Bmatrix} = \begin{Bmatrix} \mathbf{R} \\ -\mathbf{K}_{32}^{(b)}(\mathbf{K}_{22}^{(b)})^{-1}\mathbf{R}^{(b)} \end{Bmatrix}. \tag{8.58}$$

Since the bubble enriched velocity is zero at the boundary of each element, the nodal values of $\mathbf{V}^{(l)}$ are in fact the required solution. The unknown $\mathbf{V}^{(b)}$ does not need to be solved. Here $\mathbf{K}_{11}$ is a function of $\mathbf{V}^{(l)}$ due to the shear-rate dependent viscosity. This makes the system of equations non-linear. One may simply solve the equations using a Picard iteration where $\mathbf{K}_{11}$ is evaluated using the value $\mathbf{V}^{(l)}$ obtained from the previous step and so a sequence of linear problems is solved. Alternatively, some authors (e.g., Pichelin and Coupez 1998) have used the Newton-Raphson iteration.

As alternatives to the MINI element, Rajupalem et al. (1997) and Talwar et al. (1998) used an equal-order velocity–pressure finite element formulation, as proposed by Rice et al. (1986), in their 3D injection molding simulation, and Chang and Yang (2001) used a 3D finite-volume approach based on the SIMPLE finite volume algorithm of Patankar (1980). Since the finite volume method does not require large memory, it is believed to be suitable for solving large 3D problems (Tanner et al. 1995).

To solve the energy equation for the temperature field, finite element methods can also be used. The thermal conductivity of polymers is very low and hence in the flow direction the dominant heat transport mechanism is convection. As emphasized before, the classical Galerkin techniques are often unstable in solving convection-dominated problems. An upwinding scheme is the most commonly used technique to avoid the stability difficulty. Moreover, the 3D heat transfer simulation has another problem in calculating the temperature field across the thickness, where the main heat transport mechanism is thermal conduction. It is difficult to capture the rapid spatial change of the temperature along the thickness direction, especially within the narrow region next to the mold-polymer interface. This may require the use of many elements across the part thickness. However, for typical thin-walled parts, isotropic mesh refinement in the thickness direction can lead to an enormous increase in the total number of elements and prohibitive computational cost. The problem is alleviated by Gruau and Coupez (2005) who have developed a 3D tetrahedral, unstructured and anisotropic mesh generator that can place enough meshes through the thickness direction without introducing too many nodes in other directions. A different technique was presented by Friedl et al. (2004) who adapt the analytical solution for an 1D semi-infinite heat conduction to approximate the gap-wise temperature profile and hence they can retain a relatively coarse mesh across the thickness.

8.2.6 Smoothed Particle Hydrodynamics (SPH) Method

Among several numerical methods, the finite element method (FEM) is a robust and thoroughly developed method. However, the accuracy and efficiency of the method rely on the quality of meshes or elements. Usually the FEM user has to spend considerable time in mesh creation or in fixing mesh problems. Therefore, some efforts have also been made to explore the applicability of meshless methods, such as the Smoothed Particle Hydrodynamics (SPH) method.

SPH is a fully Lagrangian method to model fluid flows with complex moving boundaries and dynamic response of materials involving large deformation, fracture and fragmentation. The technique was first introduced by Gingold and Monaghan (1977) and Lucy (1977) in simulations of non-axisymmetric phenomena in astrophysics. The book of Liu and Liu (2003) gives a detailed description of SPH theory, including numerical programs and some application examples. The early implementation of the SPH method was appropriate for an inviscid fluid in gas dynamics. It has been further developed for wider applications to problems of Newtonian, generalized Newtonian, viscoelastic, and Bingham-like fluid flows, see, for example, Shao and Lo (2003), Ellero and Tanner (2005), Fang et al. (2006), Hosseini et al. (2007), Rafiee et al. (2007), Rafiee (2008), and Zhu et al. (2010). Cleary et al. (2002) and Prakash et al. (2009) used SPH to simulate the metal flow in the casting and furnace emptying with some success. An overview on the SPH method and its recent development is covered in a review paper by Liu and Liu (2010). However, application of the SPH to mold filling of polymer liquid had not been reported until the recent work by Fan et al. (Fan et al. 2010a, 2011).

In SPH, the fluid is discretized into a finite number of moving points, or "particles", where any physical quantity $f(\mathbf{x})$ associated with the particle at the position $\mathbf{x}$ is interpolated using function values at neighboring particles within a small local support domain of the position $\mathbf{x}$, i.e.,

$$f(\mathbf{x}) = \sum_{b=1}^{N} \frac{m_b}{\rho_b} f(\mathbf{x}_b) W(\mathbf{x} - \mathbf{x}_b, h), \tag{8.59}$$

where $f(\mathbf{x})$ is known only at N discrete particles at $\mathbf{x}_b (b = 1, 2, \ldots N)$, and the particle b has mass m_b and density ρ_b, and W is a kernel function and h is the smoothing length representing the effective width of the kernel. Several commonly used kernel functions can be found in the book of Liu and Liu (2003).

The SPH discretized governing equations are as follows:

The mass conservation equation:

$$\rho(\mathbf{x}_a) = \sum_{b=1}^{N} m_b W_{ab}. \tag{8.60}$$

The momentum equation:

$$\frac{d\mathbf{u}_a}{dt} = \sum_{b=1}^{N} m_b \left[-\left(\frac{P_a}{\rho_a^2} + \frac{P_b}{\rho_b^2} \right) \mathbf{I} + \left(\frac{\boldsymbol{\tau}_a}{\rho_a^2} + \frac{\boldsymbol{\tau}_b}{\rho_b^2} \right) \cdot \right] \nabla_a W_{ab} + \mathbf{b}, \tag{8.61}$$

where $\mathbf{u}$ is the velocity vector, P is the pressure, $\boldsymbol{\tau}$ is the extra stress tensor, and $\mathbf{b}$ the field force. The subscripts "a" and "b" denote particles a and b, respectively. The extra stress tensor depends on the constitutive equations used. For a specific formulation for generalized Newtonian fluids, see Fan et al. (2010a).

The energy conservation equation (Cleary and Monaghan 1999):

$$\frac{de_a}{dt} = \sum_{b=1}^{N} \frac{4m_b}{\rho_a \rho_b} \frac{k_a k_b}{(k_a + k_b)} (T_a - T_b) \frac{(\mathbf{x}_a - \mathbf{x}_b) \cdot \nabla_a W_{ab}}{r_{ab}^2 + \xi^2}, \tag{8.62}$$

where e_a is the internal energy corresponding to particle a; k_a and k_b are the thermal conductivities corresponding to particles a and b, respectively; T_a and T_b are the temperatures conductivities corresponding to particles a and b, respectively; r_{ab} is the distance between particles a and b, and ξ is a small parameter to avoid the singularity when $r_{ab} \to 0$. The equation has neglected the viscous dissipation and the heat source terms.

To simulate confined fluid flow such as mold filling, one usually models the solid boundaries by "solid particles". These particles are fixed or moving with the velocity of the moving boundaries and interact with fluid particles. Several approaches have been proposed to handle the solid boundary conditions. Most common approaches are the repulsive boundary condition method proposed by Monaghan (1992, 1994, 2005) and the ghost particle method described by Morris et al. (1997) and Ferrari et al. (2009).

The explicit integration methods, such as leapfrog, prediction-correction or Runge-Kutta methods, are usually used to integrate SPH equations for fluid flows. The explicit time integration is conditionally stable. The time step should satisfy the convective stability constraint, i.e., the so-called Courant–Friedrichs–Lewy (CFL) condition,

$$\Delta t_c \leq \alpha_c \frac{h}{u_{nax}}, \tag{8.63}$$

where u_{nax} is the maximal velocity of particles and α_c is a constant with the order of 0.1.

If the diffusion term is integrated explicitly, the time step should satisfy the diffusive stability constraint,

$$\Delta t_c \leq \alpha_d \frac{\rho h^2}{\eta_0}, \tag{8.64}$$

where α_d is the constant with the order of 0.1 and η_0 the characteristic viscosity of the fluid.

Hence, for fully explicit integration the stable time step should satisfy

$$\Delta t \leq \min(\Delta t_c, \Delta t_d). \tag{8.65}$$

For polymer flows, the viscosity is very high and the Reynolds number is very small. The diffusive stability constraint becomes determinant and the stable time step may be very small. When the time step is prohibitively small, the simulation is infeasible. In this case, implicit time integration on the diffusion term has to be used to avoid the time step limitations (Fan et al. 2010a, 2011).

In the SPH simulation, when the material is in a state of tensile stress, the particle motion may become unstable, leading to the so-called tensile instability. One can add an artificial pressure (Monaghan 2000) to stabilize the simulation.

Fan et al. (2011) used the SPH method to simulate the fountain flow phenomenon for a power-fluid. The resulting flow pattern of the fountain flow region is shown in Fig. 8.11. The initial configuration for the SPH simulation is that of a rectangular domain with a total of 4,654 particles, including 362 wall particles and 4,292 fluid particles. The repulsive boundary condition method was used. In the simulation, the fluid particles are driven by the two horizontally parallel walls moving from left to right with a velocity $\mathbf{u}_w$, but a left fixed vertical wall (not shown in the figure) blocks the flow passage. This is equivalent to setting a uniform velocity profile on the entrance boundary. Despite the simplicity of the inlet boundary condition, comparisons with finite element simulations (e.g., Mavridis et al. (1986) and Mitsoulis (2010)) are in good agreement. While the

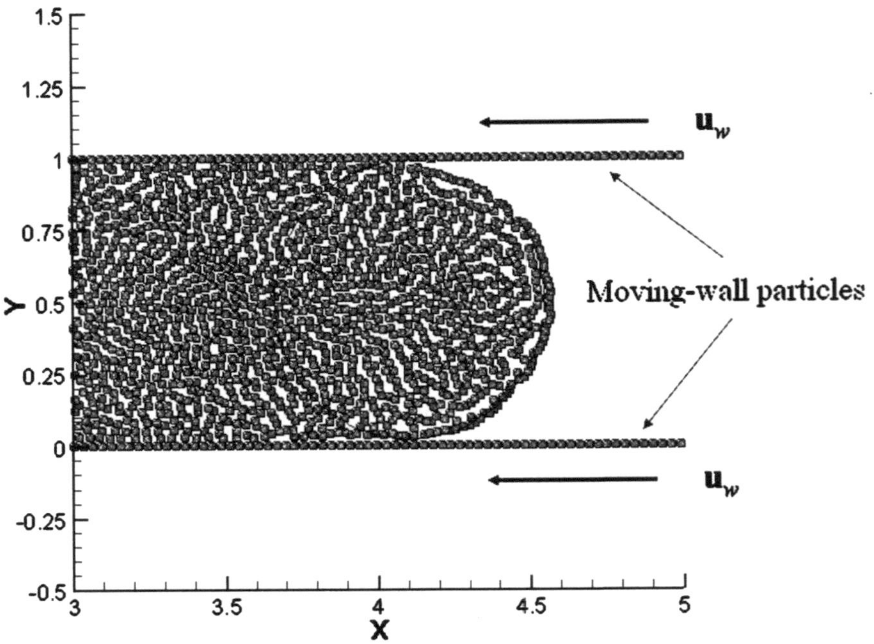

Fig. 8.11 Power-law fountain flow pattern simulated using the SPH method (Reproduced from Fan et al. (2011), with permission from Nova Science)

finite element calculation of the free surface usually requires remeshing, the SPH simulates the free surface naturally.

Compared with the grid-based methods, the SPH method handles the fountain flow more easily and naturally, without mass loss. However, for injection molding simulations, the SPH method is more time consuming, and it is difficult to enforce essential boundary conditions.

8.3 Structural Analysis for Shrinkage and Warpage Prediction

8.3.1 Shell Finite Elements

The structural analysis can be done using finite element methods. The finite element equation generated from Eq. 6.60 takes the following general form:

$$[\mathbf{K}_e]\{\mathbf{d}\} = \{\mathbf{R}\}, \tag{8.66}$$

where $\{\mathbf{d}\}$ is the unknown displacement vector, $[\mathbf{K}_e]$ is the element stiffness matrix given by

$$[\mathbf{K}_e] = \int_{V_e} [\mathbf{B}_e]^T [\mathbf{C}][\mathbf{B}_e] dV, \tag{8.67}$$

in which $[\mathbf{B}_e]$ is the so-called *strain-displacement matrix*, which operating on $\{\mathbf{d}\}$ produces strains, $\{\varepsilon\} = [\mathbf{B}_e]\{\mathbf{d}\}$, $[\mathbf{C}]$ is the matrix of material stiffness. $\{\mathbf{R}\}$ is the vector of loads formed by the thermally and pressure-induced stresses $[\sigma^0]$.

$$[\mathbf{R}] = -\int_{V_e} [\mathbf{B}_e]^T [\boldsymbol{\sigma}^0] dV. \tag{8.68}$$

After assembling Eq. 8.66 to form the global structure equations, the nodal displacement of the structure, and hence shrinkage and warpage, can be calculated. As boundary conditions, a suitable set of support constraints have to be imposed to the structure to prevent the body from undergoing unlimited rigid body motion.

Many early simulations for plate structures were based on the shell element formulations introduced by Ahmed et al. (1970). While such elements are capable of dealing satisfactorily with thick plate and shell problems, their structural response becomes too stiff as the plate or shell becomes thin. This phenomenon is called *shear locking* and is known to be related to the inability of the element to approach the limiting condition of zero transverse shear strain at the appropriate quadratic rate.

The element formulations widely used today to overcome the difficulty are triangular facet shell elements constructed by superimposing the local membrane formulation due to Bergan and Felippa (1985) with the linear bending formulation due to Batoz and Lardeur (1989) and transforming the combined equations to the global coordinate system. Each element has three nodes with 18 degrees of freedom for each element (six at each node). The material properties can vary from one element to another. Within the restriction that the material properties are not strongly asymmetric about the mid-surface, it is also possible to include variations of the mechanical properties across the thickness.

8.3.2 Dual-Domain Structural Analysis

We have already described the dual-domain approach for injection molding flow analysis. The dual domain concept has been extended to the structural analysis by Fan et al. (2000b, 2004b), in order to model the structural performance (i.e., bending and membrane strains) of a thin-walled part using only the surface mesh defining the outer boundary of the object. Geometrically, a flat-plate with thickness h can be seen as two 0.5 h-thick plates bonded together perfectly. Such an assembled structure could be modeled by using two shells, each with their reference surface at their geometric centre. However, it is more convenient to use the outer surface rather than the geometric centre as the reference surface. This can be achieved by using eccentric shell elements. The top plate uses an eccentric shell element with the top surface as its reference surface, while the bottom one uses the bottom surface as its reference surface. The bonding of the top and bottom plates involves imposing the Love-Kirchhoff assumption of classical plate or shell theory (see for example, Donnell 1976) and requires that a normal to the plate or shell remains straight after deformation and is unchanged in length. Multi-point constraints are used for this purpose.

The constraints used depend on the type of element chosen. As an example, consider the triangular facet shell element mentioned above. The element has six degrees of freedom (DOF) at each node, including three displacement DOF (u_1, u_2, u_3 and three rotational DOF (θ_1, θ_2, θ_3), defined in the local element system. The drilling rotational DOF about a local reference surface normal is used in the membrane formulation, and is given by

$$\theta_3 = \frac{1}{2}\left[\frac{\partial u_2}{\partial x_1} - \frac{\partial u_1}{\partial x_2}\right], \tag{8.69}$$

Assume that a node n at the bottom element matches the node p at the top element as shown in Fig. 8.12a. A node is said to match a point on the opposite surface, if their connection line is parallel to the normal of the midsurface. We require that the normal to the midsurface before deformation should remain straight after deformation. Then there is a relationship between the degrees of freedom of node n and point p, which can be written as

$$\left\{\mathbf{U}^{(n)}\right\} = [\mathbf{A}]\left\{\mathbf{U}^{(p)}\right\}, \tag{8.70}$$

where $\left\{\mathbf{U}^{(n)}\right\}$ and $\left\{\mathbf{U}^{(p)}\right\}$denote the DOF on nodes n and p respectively.

$$\left\{\mathbf{U}^{(n)}\right\} = \left\{u_1^{(n)}, u_2^{(n)}, u_3^{(n)}, \theta_1^{(n)}, \theta_2^{(n)}, \theta_3^{(n)}\right\}^T, \tag{8.71}$$

$$\left\{\mathbf{U}^{(p)}\right\} = \left\{u_1^{(p)}, u_2^{(p)}, u_3^{(p)}, \theta_1^{(p)}, \theta_2^{(p)}, \theta_3^{(p)}\right\}^T, \tag{8.72}$$

and

$$[\mathbf{A}] = \begin{bmatrix} 1 & 0 & 0 & 0 & -h & 0 \\ 0 & 1 & 0 & h & 0 & 0 \\ 0 & 0 & 1 & 0 & 0 & 0 \\ 0 & 0 & 0 & 1 & 0 & 0 \\ 0 & 0 & 0 & 0 & 1 & 0 \\ 0 & 0 & 0 & \frac{h}{2}\frac{\partial}{\partial x_1} & \frac{h}{2}\frac{\partial}{\partial x_2} & 1 \end{bmatrix}, \tag{8.73}$$

where h is the distance between node n and its matching point p. In general, the mesh on the top surface is not coincident with the bottom mesh. It is more likely that the matching point p will be an interior point of a top element as shown in Fig. 8.12b. In this case, one interpolates the required constraints using the three nodes defining the top element.

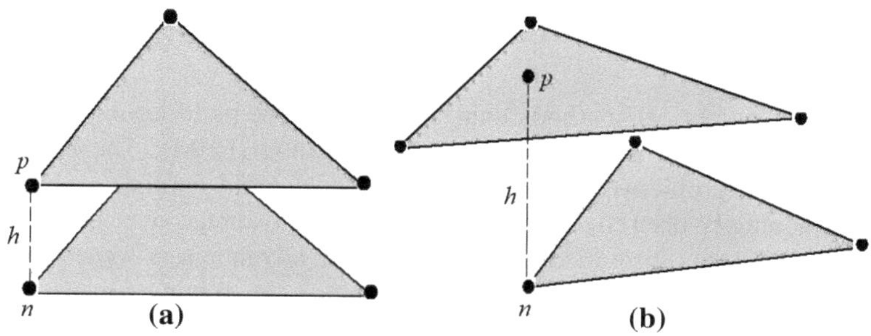

Fig. 8.12 The matching points in dual domain structural analysis. **a** Node n at the bottom element matches a node at the *top element*. **b** Node n at the bottom element matches a point within the *top element*

This system of constraints is imposed at all nodes on the bottom (or top) surface of the model. With these constraints, the structural performance of the composite structure is identical to the original plate.

8.3.3 3D Structural Analysis

Warpage analysis using three-dimensional elements has been discussed by Fan et al. (2004a). If a 4-node first-order tetrahedral element is used for the warpage analysis of typical thin-walled parts, a shear locking problem will occur that makes the structural response very stiff, especially for elements with large aspect ratios. For a tetrahedral element, the aspect ratio is defined as the ratio of the largest dimension to the diameter of the largest sphere that can be inscribed in the element. In thin-wall regions of a part, high aspect ratio tetrahedral elements are usually hard to avoid.

Fan et al. (2004a) described a hybrid element scheme for the 3D warpage analysis. The idea is to use 4-node first order tetrahedral elements only in chunky areas, but use 10-node second-order tetrahedral elements in the thin-walled areas and 5–9 node tetrahedral elements in the transitional areas.

However, since most injection-molded parts are dominated by the thin-walled structure, the elemental upgrade generally increases the total number of nodes by around 6 times. Thus the method is very demanding as far as memory requirement and computation time are concerned. Fan et al. (2006) proposed to use a parallelized algebraic multigrid preconditioned conjugate gradient (AMG-CG) solver and a dual mesh approach. The idea of the dual mesh approach is to generate a coarse mesh for the warpage analysis from the dense mesh used in the flow analysis. Multigrid mapping algorithm is used to map the results of the flow analysis (i.e., orientation, mechanical properties and thermal stresses or thermal strains) from the dense mesh to the coarse mesh, and then, after having completed the warpage analysis on the coarse mesh, map warpage results back to the dense mesh.

8.4 Boundary Element Method for Mold Cooling Analysis

While several numerical methods can be used for the mold cooling analysis, a widely used method is the boundary element method (BEM). The BEM, when applied to linear problems, such as the steady-state mold cooling problem, only requires a boundary discretization. This is a major advantage over other methods requiring full domain discretization. However, the advantage is somewhat lost in solving the transient mold cooling problem, since the inhomogeneity resulting from the time derivative term will be included in the formulation through domain integrals. In this case, finite element methods could be preferable. Still, several techniques have been developed to make the BEM applicable to non-linear and time-dependent problems. Some of the methods divide the domain into a number of domain sells for integrating numerically the domain integral term (see, for example, Bush and Tanner 1990). This method does not add new unknowns to the problem, i.e., the unknowns still refer only to the boundary. However, the large number of operations necessary in the numerical integration procedure can be costly. Alternatively, some other methods transform the domain integrals to the boundary in order to eliminate the need for internal cells. We shall briefly review some of the techniques in the following section.

8.4.1 Transient Mold Cooling

8.4.1.1 Coupled BE/FD Method

The initial-boundary value problem represented by Eq. 7.1 can be transmuted into a boundary integral equation by several different methods. Brebbia and Walker (1980) and Curran et al. (1980) approximated the time derivative in the equation in a finite difference form, thus changing the original parabolic partial differential equation to an elliptical partial differential equation, for which the standard boundary integral equation may be established.

Assuming that the mold temperature field at time t_{n-1} is known to be $T_{n-1}(\mathbf{r})$, where $\mathbf{r}$ is the position vector, then for a sufficiently small time increment Δt, Eq. 7.1 can be approximated as

$$\nabla^2 T^n(\mathbf{r}) = \frac{1}{\alpha_M}\left(\frac{T^n(\mathbf{r}) - T^{n-1}(\mathbf{r})}{\Delta t}\right). \tag{8.74}$$

where $T^n(\mathbf{r})$ is the unknown temperature field at $t_n = t_{n-1} + \Delta t$. The equation is then written as

$$\nabla^2 T^n(\mathbf{r}) - \frac{1}{\alpha_M \Delta t} T^n(\mathbf{r}) = -\frac{1}{\alpha_M \Delta t} T^{n-1}(\mathbf{r}). \tag{8.75}$$

The type of Eq. 8.75 is elliptical and the boundary integral equation for the formulation can be derived to give

$$c(\mathbf{r}_0)T^n(\mathbf{r}_0) + \alpha_M \int_\Gamma \left(T^n(\mathbf{r}) \frac{\partial T^*(\mathbf{r}, \mathbf{r}_0, \Delta t)}{\partial n} - T^*(\mathbf{r}, \mathbf{r}_0, \Delta t) \frac{\partial T^n(\mathbf{r})}{\partial n} \right) d\Gamma(\mathbf{r})$$
$$= \frac{1}{\Delta t} \int_\Omega T^*(\mathbf{r}, \mathbf{r}_0, \Delta t) T^{n-1}(\mathbf{r}) d\Omega(\mathbf{r}), \tag{8.76}$$

where $\mathbf{r}_0$ is the load point and $\mathbf{r}$ is the field point,

$$c(\mathbf{r}_0) = \begin{cases} 0.5 & \text{if} \quad \mathbf{r}_0 \in \Gamma \text{ and } \Gamma \text{ is smooth at } \mathbf{r}_0 \\ 1 & \text{if} \quad \mathbf{r}_0 \in \Omega - \Gamma \end{cases}, \tag{8.77}$$

and $T^*(\mathbf{r}, \mathbf{r}_0, \Delta t)$ is the fundamental solution satisfies

$$\alpha_M \nabla^2 T^* - \frac{1}{\Delta t} T^* = \delta(\mathbf{r} - \mathbf{r}_0), \tag{8.78}$$

with δ being the Dirac delta function. The fundamental solution T^* for the 3-D problem is given by

$$T^* = \frac{(\alpha_M/\Delta t)^{1/4}}{|\mathbf{r} - \mathbf{r}_0|^{1/2} (2\pi\alpha_M)^{3/2}} K_{1/2} \left(\frac{|\mathbf{r} - \mathbf{r}_0|}{(\alpha_M \Delta t)^{1/2}} \right), \tag{8.79}$$

where $K_{1/2}(z)$ is the modified Bessel function of the second kind. The limiting form of $K_{1/2}(z)$ as $z \to 0$ is $K_{1/2}(z) = [\pi/(2z)]^{1/2}$ (Antosiewitcz 1972).

Starting from a given initial temperature field T at $t_{n-1} = t_0$, one can solve Eq. 8.76 for the unknown temperature field at the next time step and advance the solution in time. The domain integral term has to be evaluated at each time level because T^{n-1} changes at each time step.

8.4.1.2 BIE/Particular Solution Method

In the coupled BIE/FD formulation mentioned above, the requirement for the evaluation of the domain integral is certainly a costly feature. Zheng and Phan-Thien (1992), who considered a boundary integral equation for the following problem of heat conduction with a heat source, have employed an alternative approach that transforms the domain integral to a boundary integral.

$$\nabla^2 T = f(T), \quad \text{in } \Omega \text{ with boundary } \Gamma, \tag{8.80}$$

where $f(T)$ represents the heat source. Obviously, Eq. 8.74 belongs to the class of problem which considers the time derivative term as a pseudo heat source.

In Eq. 8.80, $f(T)$ is not known explicitly, because it depends on the unknown variable T. However, with Picard iteration procedure the problem can be reduced to that of solving a sequence of Poisson equations:

$$\nabla^2 T = f(\mathbf{r}), \quad \text{in } \Omega \text{ with boundary } \Gamma. \tag{8.81}$$

An alternative to the iteration method is to take the unknown terms in $f(T)$ to the left-hand side of (8.80), and thus the remaining terms on the right-hand side of the equation are known from the previous time step, as expressed by Eq. 8.75. In this case, the problem will be reduced to solving a sequence of modified Bessel equations (see Ingber and Phan-Thien 1992 for more details).

The integral equation for (8.81) is then given by

$$c(\mathbf{r}_0)T(\mathbf{r}_0) + \int_\Gamma \left(T(\mathbf{r}) \frac{\partial T^*(\mathbf{r}, \mathbf{r}_0)}{\partial n} - T^*(\mathbf{r}, \mathbf{r}_0) \frac{\partial T(\mathbf{r})}{\partial n} \right) d\Gamma(\mathbf{r}) = - \int_\Omega T^*(\mathbf{r}, \mathbf{r}_0) f(\mathbf{r}) d\Omega(\mathbf{r}), \tag{8.82}$$

where

$$c(\mathbf{r}_0) = \begin{cases} 0.5 & \text{if} \quad \mathbf{r}_0 \in \Gamma \text{ and } \Gamma \text{ is smooth at } \mathbf{r}_0 \\ 1 & \text{if} \quad \mathbf{r}_0 \in \Omega - \Gamma \end{cases}, \tag{8.83}$$

and T^* satisfies

$$\nabla^2 T^* = \delta(\mathbf{r} - \mathbf{r}_0), \tag{8.84}$$

and for 3D problems,

$$T^* = \frac{1}{4\pi |\mathbf{r} - \mathbf{r}_0|}. \tag{8.85}$$

A way of eliminating the domain integral in Eq. 8.81 is based on the well-known idea of combining a particular solution to a complementary solution of a homogeneous problem. This idea was discussed in Buzbee et al. (1971), who named the method the *capacitance matrix method*. Brebbia and Nardini (1986) have applied the concept to boundary element method and called it the *dual reciprocity method* (DRM). Henry and Banerjee (1988) have also proposed a related method in the name of the *particular integral method*. Now, let us consider a particular solution $\hat{T}$ that satisfies

$$\nabla^2 \hat{T} = f(\mathbf{r}), \tag{8.86}$$

To find a particular solution is not difficult, since it can be any solution satisfying the equation, without having to consider the geometry or any boundary conditions of the problem. Of course, the particular solution is not unique. Although closed-form particular solutions are not always available, approximate particular solutions can be derived in several ways, and the simplest approach is to represent the inhomogeneity in terms of simpler functions for which the solution is known. Zheng and Phan-Thien (1992) used the radial basis function (RBF) interpolation, following the earlier work of Coleman (1990), Zheng et al. (1991a, b, 1992) and Coleman et al. (1991). With this method, the inhomogeneity f is replaced by $f(\mathbf{r}) \approx \sum_{i=1}^{N} \alpha_i \exp\left[-(|\mathbf{r}-\mathbf{r}_i|/\beta_i)^2\right]$, where the $\mathbf{r}_i'$s are chosen to be points distributed within the interior domain, and the β_i's are chosen to be the average distance of the adjacent nodes to $\mathbf{r}_i$. Those choices of the parameters leads to a good interior approximation for $f(\mathbf{r})$, which can be determined from a sparse system of equations for the α_i's. Thus an approximate particular solution of the governing Eq. 8.86 can be derived.

Using the particular solution for Eq. 8.86, the volume integral term in Eq. 8.82 becomes

$$\int_{\Omega} T^* f d\Omega = \int_{\Omega} T^* \nabla^2 \hat{T} d\Omega. \tag{8.87}$$

Considering Green's second identity and using Eq. 8.84, we then have

$$\int_{\Omega} T^* f d\Omega = -c(\mathbf{r}_0)\hat{T}(\mathbf{r}_0) - \int_{\Gamma} \left(\hat{T}(\mathbf{r})\frac{\partial T^*(\mathbf{r},\mathbf{r}_0)}{\partial n} - T^*(\mathbf{r},\mathbf{r}_0)\frac{\partial \hat{T}(\mathbf{r})}{\partial n}\right) d\Gamma(\mathbf{r}). \tag{8.88}$$

Substituting Eq. 8.88 into 8.82 leads to

$$c(\mathbf{r}_0)T^h(\mathbf{r}_0) + \int_{\Gamma} \left(T^h(\mathbf{r})\frac{\partial T^*(\mathbf{r},\mathbf{r}_0)}{\partial n} - T^*(\mathbf{r},\mathbf{r}_0)\frac{\partial T^h(\mathbf{r})}{\partial n}\right) \mathrm{d}\Gamma(\mathbf{r}) = 0, \tag{8.89}$$

where $T^h = T - \hat{T}$. The boundary conditions for T^h are:

$$k_M \frac{\partial T^h}{\partial n} = -q - k_M \frac{\partial \hat{T}}{\partial n}, \quad \text{in } \Gamma_p, \tag{8.90}$$

$$k_M \frac{\partial T^h}{\partial n} = -h_c\left(T^h + \hat{T} - T_b\right) - k_M \frac{\partial \hat{T}}{\partial n}, \quad \text{in } \Gamma_c, \tag{8.91}$$

$$k_M \frac{\partial T^h}{\partial n} = -h_a\left(T^h + \hat{T} - T_a\right) - k_M \frac{\partial \hat{T}}{\partial n}, \quad \text{in } \Gamma_e. \tag{8.92}$$

Equation 8.89 is the desired boundary integral equation. The domain integral has been eliminated and a boundary-only formulation is obtained. After solving the equation for T^h (we call it the *homogeneous solution*), the total solution can be obtained by$T = T^h + \hat{T}$. Although the particular solution $\hat{T}$ is not unique, the total solution T obtained in this way is unique, since the equation for T^h satisfies the above modified boundary conditions.

Once the boundary values have been obtained, to calculate values in the interior is straightforward. While this method does not require domain discretization, values of T at sufficient internal points are required to update f for the next time step.

Numerical results of mold cooling analysis using a similar method were also presented by Chen and Chung (1994).

8.4.1.3 Laplace Transform

The Laplace-transform based boundary integral equation was proposed by Rizzo and Shippy (1970) to solve transient heat conduction problems.

By taking the Laplace transform of Eq. 7.1, one obtains

$$\nabla^2 \tilde{T} - \frac{1}{\alpha_M} s\tilde{T} = -\frac{1}{\alpha_M} T_0, \tag{8.93}$$

where T_0 is the initial condition, and $\tilde{T}$ is the transformed temperature

$$\tilde{T} = L(T) = \int_0^\infty T(\mathbf{r}, t) \exp(-st) dt. \tag{8.94}$$

The boundary condition, Eqs. 7.2–7.4 must also be transformed.

$$k_M \frac{\partial \tilde{T}}{\partial n} = -\tilde{q}, \quad \text{in } \Gamma_p, \tag{8.95}$$

$$k_M \frac{\partial \tilde{T}}{\partial n} = -h_c \left(\tilde{T} - \frac{T_b}{s} \right), \quad \text{in } \Gamma_c, \tag{8.96}$$

$$k_M \frac{\partial \tilde{T}}{\partial n} = -h_a \left(\tilde{T} - \frac{T_a}{s} \right), \quad \text{in } \Gamma_e. \tag{8.97}$$

Equation 8.93 is an elliptical partial differential equation, and it is similar in form to Eq. 8.75, so that its fundamental solutions have the same form as Eq. 8.79, with Δt replaced by $1/s$. The boundary integral equation for Eq. 8.93 is

$$c(\mathbf{r}_0)\tilde{T}(\mathbf{r}_0, s) + \alpha_M \int_\Gamma \left(\tilde{T}(\mathbf{r}, s) \frac{\partial \tilde{T}^*(\mathbf{r}, \mathbf{r}_0, s)}{\partial n} - \tilde{T}^*(\mathbf{r}, \mathbf{r}_0, s) \frac{\partial \tilde{T}(\mathbf{r}, s)}{\partial n} \right) d\Gamma(\mathbf{r})$$
$$= \int_\Omega \tilde{T}^*(\mathbf{r}, \mathbf{r}_0, s) T_0(\mathbf{r}) d\Omega(\mathbf{r}) \tag{8.98}$$

After the solution is obtained in the Laplace space, a numerical transform inversion $T(\mathbf{r}, t) = L^{-1}[\tilde{T}(\mathbf{r}, s)]$ is then performed to convert the solution to the original time domain.

8.4.1.4 Time-Dependent Fundamental Solution

Another alternative BIE that can be applied to mold cooling analysis is to use time-dependent fundamental solutions (e.g., Qiao 2005). The boundary integral equation can be written as

$$c(\mathbf{r}_0)T(\mathbf{r}_0,t) + \alpha_M \int_0^t \int_\Gamma \left(T(\mathbf{r},t') \frac{\partial T^*(\mathbf{r},\mathbf{r}_0,t,t')}{\partial n} - T^*(\mathbf{r},\mathbf{r}_0,t,t') \frac{\partial T(\mathbf{r},t')}{\partial n} \right) \mathrm{d}\Gamma(\mathbf{r})\mathrm{d}t'$$
$$= \int_\Omega T^*(\mathbf{r},\mathbf{r}_0,t,t')T_0(\mathbf{r})\mathrm{d}\Omega(\mathbf{r}), \tag{8.99}$$

where

$$T^* = \frac{1}{[4\pi\alpha_M(t-t')]^{3/2}} \exp\left[-\frac{|\mathbf{r}-\mathbf{r}_0|^2}{4\alpha_M(t-t')} \right], \quad \text{for 3D problems.} \tag{8.100}$$

8.4.2 Steady-State Mold Cooling

As has been mentioned in Sect. 7.3, the continuous injection-molding operation results in a cyclic heat transfer behavior in the mold, after a short transient period. The cycle-averaged temperature can be represented by a steady state heat conduction equation, i.e., Eq. 7.10. The mold cooling analysis can be greatly simplified by solving the steady state problem. The boundary integral equation of (7.10) is

$$c(\mathbf{r}_0)T(\mathbf{r}_0) + \int_\Gamma \left(T(\mathbf{r}) \frac{\partial T^*(\mathbf{r},\mathbf{r}_0)}{\partial n} - T^*(\mathbf{r},\mathbf{r}_0) \frac{\partial T(\mathbf{r})}{\partial n} \right) d\Gamma(\mathbf{r}) = 0. \tag{8.101}$$

where

$$c(\mathbf{r}_0) = \begin{cases} 0.5 & \text{if} \quad \mathbf{r}_0 \in \Gamma \text{ and } \Gamma \text{ is smooth at } \mathbf{r}_0 \\ 1 & \text{if} \quad \mathbf{r}_0 \in \Omega - \Gamma \end{cases}, \tag{8.102}$$

and T^* satisfies

$$\nabla^2 T^* = \delta(\mathbf{r}-\mathbf{r}_0), \tag{8.103}$$

and for 3-D problems,

$$T^* = \frac{1}{4\pi|\mathbf{r}-\mathbf{r}_0|}. \tag{8.104}$$

8.4.3 Modified Boundary Integral Equations for Closely Spaced Surfaces

For a thin-walled molding, the distance between the opposing cavity surfaces (i.e., part thickness) is typically orders of magnitude smaller than the length and width of the part. In this case, the traditional boundary integral formulation leads to a poorly conditioned problem, and hence a modification to the boundary integral equation as described by Rezayat and Burton (1990) needs to be used. The basic idea of the modification is to merge the two opposing surfaces to a single mid-surface and treat it as an internal crack in the mold.

Associated with the midsurface, each variable has two different variables at "top" and "bottom" defined by a normal vector $\boldsymbol{n}$ to the mid-surface, one can use superscript "+" and "−" on the variables (e.g., T^+ and T^-) to indicate values on the different sides. An extra formula is thus derived for the additional degree of freedom. For point $\mathbf{r}_0$ on the midsurface, the following two boundary integral equations are solved:

$$\begin{aligned}
&\frac{1}{2}(T^+(\mathbf{r}_0)+T^-(\mathbf{r}_0)) \\
&\quad+\int_{\Gamma_p}\left\{[T^+(\mathbf{r})-T^-(\mathbf{r})]\frac{\partial T^*(\mathbf{r},\mathbf{r}_0)}{\partial n}-T^*(\mathbf{r},\mathbf{r}_0)\left[\frac{\partial T^+(\mathbf{r})}{\partial n^+}+\frac{\partial T^-(\mathbf{r})}{\partial n^-}\right]\right\}d\Gamma(\mathbf{r}) \\
&\quad+\int_{\Gamma-\Gamma_p}\left(T(\mathbf{r})\frac{\partial T^*(\mathbf{r},\mathbf{r}_0)}{\partial n}-T^*(\mathbf{r},\mathbf{r}_0)\frac{\partial T(\mathbf{r})}{\partial n}\right)d\Gamma(\mathbf{r})=0, \qquad (8.105)
\end{aligned}$$

$$\begin{aligned}
&\frac{1}{2}\left(\frac{\partial T^+(\mathbf{r}_0)}{\partial m}-\frac{\partial T^-(\mathbf{r}_0)}{\partial m}\right) \\
&\quad+\int_{\Gamma_p}\left\{\frac{\partial^2 T^*(\mathbf{r},\mathbf{r}_0)}{\partial n\partial m}[T^+(\mathbf{r})-T^-(\mathbf{r})]-\frac{\partial T^*(\mathbf{r},\mathbf{r}_0)}{\partial m}\left[\frac{\partial T^+(\mathbf{r})}{\partial n^+}+\frac{\partial T^-(\mathbf{r})}{\partial n^-}\right]\right\}d\Gamma(\mathbf{r}) \\
&\quad+\int_{\Gamma-\Gamma_p}\left(\frac{\partial^2 T^*(\mathbf{r},\mathbf{r}_0)}{\partial n\partial m}T(\mathbf{r})-\frac{\partial T^*(\mathbf{r},\mathbf{r}_0)}{\partial m}\frac{\partial T(\mathbf{r})}{\partial n}\right)d\Gamma(\mathbf{r})=0, \qquad (8.106)
\end{aligned}$$

where $\boldsymbol{m}$ is the outward unit normal at point $\mathbf{r}_0$.

For a point $\mathbf{r}_0$ on the exterior surface of the mold, Γ_e, one has

$$\begin{aligned}
&\frac{1}{2}T(\mathbf{r}_0)+\int_{\Gamma_p}\left\{[T^+(\mathbf{r})-T^-(\mathbf{r})]\frac{\partial T^*(\mathbf{r},\mathbf{r}_0)}{\partial n}-T^*(\mathbf{r},\mathbf{r}_0)\left[\frac{\partial T^+(\mathbf{r})}{\partial n^+}+\frac{\partial T^-(\mathbf{r})}{\partial n^-}\right]\right\}d\Gamma(\mathbf{r}) \\
&\quad+\int_{\Gamma-\Gamma_p}\left[T(\mathbf{r})\frac{\partial T^*(\mathbf{r},\mathbf{r}_0)}{\partial n}-T^*(\mathbf{r},\mathbf{r}_0)\frac{\partial T(\mathbf{r})}{\partial n}\right]d\Gamma(\mathbf{r})=0, \qquad (8.107)
\end{aligned}$$

The cooling channel surfaces have been divided into N cylindrical segments. For a point $\mathbf{r}_0$ on the axis of the jth cylindrical segment of the cooling channel, one has

$$\int_{l_j}\int_{\Gamma_p}\left\{[T^+(\mathbf{r})-T^-(\mathbf{r})]\frac{\partial T^*(\mathbf{r},\mathbf{r}_0)}{\partial n}-T^*(\mathbf{r},\mathbf{r}_0)\left[\frac{\partial T^+(\mathbf{r})}{\partial n^+}+\frac{\partial T^-(\mathbf{r})}{\partial n^-}\right]\right\}d\Gamma(\mathbf{r})dl(\mathbf{r}_0)$$
$$+\int_{l_j}\int_{\Gamma_e}\left[T(\mathbf{r})\frac{\partial T^*(\mathbf{r},\mathbf{r}_0)}{\partial n}-T^*(\mathbf{r},\mathbf{r}_0)\frac{\partial T(\mathbf{r})}{\partial n}\right]d\Gamma(\mathbf{r})dl(\mathbf{r}_0)$$
$$+\int_{l_j}\sum_{n=1}^{N}\int_{l_n}\int_{0}^{2\pi}\left[T(\mathbf{r})\frac{\partial T^*(\mathbf{r},\mathbf{r}_0)}{\partial n}-T^*(\mathbf{r},\mathbf{r}_0)\frac{\partial T(\mathbf{r})}{\partial n}\right]R_j d\theta dl(\mathbf{r}_0)dl(\mathbf{r})=0 \quad (8.108)$$

where l and θ are the axial and azimuthal coordinates, respectively, and R_j is the radius of jth cooling channel segments. Here the tip surfaces of the cooling channel have been neglected. For further consideration of the tip surfaces, see Park and Kwon (1998).

8.4.4 Boundary Discretization

Having defined the boundary integral equations, next step is to discretize the integral equations and find solutions. Let us use Eq. 7.40 as an example. After substituting the boundary conditions on the mold cavity surface (Γ_p), the cooling channel surface (Γ_c) and the external surface (Γ_e), respectively, the equation can be re-written as

$$cT+\int_{\Gamma_p}\left(T\frac{\partial T^*}{\partial n}+T^*\frac{1}{k_M}\bar{q}\right)d\Gamma+\int_{\Gamma_c}\left(T\frac{\partial T^*}{\partial n}+T^*\frac{h_c}{k_M}(T-T_b)\right)d\Gamma$$
$$+\int_{\Gamma_e}\left(T\frac{\partial T^*}{\partial n}+T^*\frac{h_c}{k_M}(T-T_a)\right)d\Gamma=0. \quad (8.109)$$

The entire boundary $\Gamma=\Gamma_p+\Gamma_c+\Gamma_e$ is then divided into a total of N non-overlapped elements, with N_p elements on Γ_p, numbered from 1 to N_p, N_c elements on Γ_c, numbered from N_p+1 to N_p+N_c, N_e elements on Γ_e, numbered from N_p+N_c+1 to N, collectively written as $\Gamma_p=\cup_{i=1}^{N_p}e_i$, $\Gamma_c=\cup_{i=N_p+1}^{N_p+N_c}e_i$, and $\Gamma_e=\cup_{i=N_p+N_c+1}^{N}e_i$, where $\mathbf{e}_i$ stands for the ith element. The variables are assumed uniform over each element, and the nodes are located at the centroid of each element. Then Eq. 8.109 is discretized to a set of linear equations

$$\frac{1}{2}T_i + \sum_{j=1}^{N_p}\left[\int_{e_j}\frac{\partial T^*_{ij}}{\partial n}dS\right]T_j + \sum_{j=N_p+1}^{N_p+N_c}\left[\int_{e_j}\left(\frac{\partial T^*_{ij}}{\partial n}+\frac{h_c}{k_M}T^*_{ij}\right)dS\right]T_j$$
$$+\sum_{j=N_p+N_c+1}^{N}\left[\int_{e_j}\left(\frac{\partial T^*_{ij}}{\partial n}+\frac{h_e}{k_M}T^*_{ij}\right)dS\right]T_j + \sum_{j=1}^{N_p}\left[\int_{e_j}\frac{h_e}{k_M}T^*_{ij}dS\right]\bar{q}_j$$
$$-\sum_{j=N_p+1}^{N_p+N_c}\left[\int_{e_j}\frac{h_c}{k_M}T^*_{ij}dS\right](T_b)_j - \left[\sum_{j=N_p+N_c+1}^{N}\left(\int_{e_j}T^*_{ij}dS\right)\right]\frac{h_c}{k_M}T_a$$
$$= 0, (i = 1, 2, \ldots N), \tag{8.110}$$

where T_j denotes the nodal temperature at element e_j. The ½-factor arises because point i is located on the boundary. There are N unknown components of T in N equations. The equations then take the following matrix form:

$$[\mathbf{A}]\{\mathbf{T}\} = \{\mathbf{B}\}, \tag{8.111}$$

where $\{\mathbf{T}\}$ is the vector of unknown variables, $[\mathbf{A}]$ is the system stiffness matrix, resulting from the integration of the fundamental solutions, and $\{\mathbf{B}\}$is the right-hand side vector, resulting from the boundary conditions and depending on the integration of the fundamental solutions. This system of equations is diagonally dominant, and is said to be well-posed. However, the matrix $[\mathbf{A}]$ is fully populated, resulting from the fact that all elements interact with all the other. This implies that for problems with N unknowns the memory storage and the number of operations to solve the system of equations with an iterative solver are $O(N)^2$. To accelerate the BEM computation, we can split the matrix into a near-field part due to elements that are close to each other and a far-field part for the remaining elements, i.e.,

$$[\mathbf{A}] = [\mathbf{A}_{\text{near}}] + [\mathbf{A}_{\text{far}}]. \tag{8.112}$$

Thus, Eq. 8.111 can be written as

$$[\mathbf{A}_{\text{near}}]\{\mathbf{T}\} = [\mathbf{B}] - [\mathbf{A}_{\text{far}}]\{\mathbf{T}\}. \tag{8.113}$$

Since for each element, there are only a small number of near-field elements, the matrix $[\mathbf{A}_{\text{near}}]$ will be sparse. The far-field matrix $[\mathbf{A}_{\text{far}}]$ is still dense, but the values of its components are relatively small, because of the $O\left(|\mathbf{r}-\mathbf{r}_0|^{-1}\right)$ behavior of T^* and the $O\left(|\mathbf{r}-\mathbf{r}_0|^{-2}\right)$ behavior of $\partial T^*/\partial n$. Considering that $[\mathbf{B}]$ (which includes the unknown boundary condition $\bar{q}$) has to be updated iteratively, the product $[\mathbf{A}_{\text{far}}]\{\mathbf{T}\}$ can be updated at the same time using the value of T at the previous iteration. This provides a good approximation, without increasing the iteration number. In the calculation of $[\mathbf{A}_{\text{far}}]\{\mathbf{T}\}$, elements far away from the collocation point $\mathbf{r}_0$ can be aggregated in order to further speed up the computation. In addition, when the collocation point is far from the integrating surface, a method called the *fast multipole method* (FMM) can be effectively employed to reduce the integrating time. As an example, consider an integral of the type $\int_S G_{ij}(\mathbf{r}_0, \mathbf{r})T_j(\mathbf{r})dS(\mathbf{r})$, where $G_{ij}(\mathbf{r}_0, \mathbf{r})$ can either be $T^*(\mathbf{r}_0, \mathbf{r})$ or $\partial T^*(\mathbf{r}_0, \mathbf{r})/\partial n$, and S is either an element or a surface consisting of a group of elements. This integration has to be calculated each time when the collocation point $\mathbf{r}_0$ is changed. Let $\mathbf{r}_c$ be a point on S. When the surface S. is far away from the collocation point $\mathbf{r}_0$, i.e., $|\mathbf{r}_0 - \mathbf{r}_c| \gg |\mathbf{r} - \mathbf{r}_c|$, the integration above can be evaluated approximately from its truncated multipole expansion. One can write

$$G_{ij}(\mathbf{r}_0, \mathbf{r}) = G_{ij}(\mathbf{r}_0, \mathbf{r}_c) + (\mathbf{r} - \mathbf{r}_c) \cdot \nabla G_{ij}(\mathbf{r}_0, \mathbf{r}_c) + \ldots, \tag{8.114}$$

Suppose that only the first term in the expansion is kept, the above integration becomes, approximately

$$\int_S G_{ij}(\mathbf{r}_0, \mathbf{r})T_j(\mathbf{r})dS(\mathbf{r}) \approx G_{ij}(\mathbf{r}_0, \mathbf{r}_c)Q_j^{(S)}, \tag{8.115}$$

where

$$Q_j^{(S)} = \int_S T_j(\mathbf{r})\mathrm{d}S(\mathbf{r}). \tag{8.116}$$

The right-hand side of Eq. 8.116 shows that $\mathbf{r}$ and $\mathbf{r}_0$ are separated due to the introduction of point $\mathbf{r}_c$, so that the integration $Q_j^{(S)}$ with respect to $\mathbf{r}$ is independent of the collocation point $\mathbf{r}_0$. This implies that such integrations only need to be computed once and can be reused when the collocation point $\mathbf{r}_0$ changed. The reader is referred to Liu (2009) for details on the FMM and its implementation for boundary element methods.

8.5 Overall Conclusion

The foregoing text shows that injection molding process predictions involve the use of a huge set of ideas from rheology and computational mechanics. We have explored the physics of the materials used (Chaps. 2, 3, 4, 5 and 6), the thermal problems (Chap. 7) and finally the computational techniques that have so far proved to be useful (this chapter).

We believe that further investigations remain to be done; improvements in the material descriptions (including commercially feasible methods of material characterization) are needed and, to a lesser extent perhaps, improvements in computational methods (including meshless methods) would be welcome. A longer goal that needs to be worked towards is the quantitative prediction of material structure and properties after molding. Whilst some progress has been made towards this end, we believe that many interesting problems and investigations lie ahead.

Appendix A

A.1 Einstein Summation Convention

If an index occurs twice in any one term, the repeated index implies a summation of that term over the range of the index. For example,

$$u_k \frac{\partial \rho}{\partial x_k} = u_1 \frac{\partial \rho}{\partial x_1} + u_2 \frac{\partial \rho}{\partial x_2} + u_3 \frac{\partial \rho}{\partial x_3}, \tag{A.1}$$

$$\sigma_{ij} n_j = \sigma_{i1} n_1 + \sigma_{i2} n_2 + \sigma_{i3} n_3. \tag{A.2}$$

A.2 The Kronecker Delta and the Alternating Tensor

The *Kronecker delta* is defined as

$$\delta_{ij} = \begin{cases} 0 & \text{if} \quad i \neq j \\ 1 & \text{if} \quad i = j \end{cases}, \tag{A.3}$$

which is also called the unit tensor, denoted by **I** in the boldface notation. Using the above definition and the summation convention one obtains

$$\delta_{ii} = 3, \tag{A.4}$$

$$\delta_{ij} \delta_{ij} = 3, \tag{A.5}$$

and

$$\delta_{ij} \delta_{ik} \delta_{jk} = 3. \tag{A.6}$$

We also have

$$u_i \delta_{ij} = u_j. \tag{A.7}$$

Here the subscript index i in u_i is replaced by j, hence the Kronecker delta is also called the *substitution tensor*.

R. Zheng et al., *Injection Molding*, DOI: 10.1007/978-3-642-21263-5,

Similarly, for a tensor A_{ij}, one has

$$A_{ij}\delta_{jk} = A_{ik}. \tag{A.8}$$

The *alternating tensor*—also called the *permutation symbol*—is defined as

$$\varepsilon_{ijk} = \begin{cases} +1 & \text{if} \quad ijk = 123,\ 231 \text{ or } 312 \\ -1 & \text{if} \quad ijk = 321,\ 132 \text{ or } 213 \\ 0 & \text{if} \quad \text{any two indices are alike} \end{cases}, \tag{A.9}$$

which satisfies the following identities:

$$\delta_{ij}\varepsilon_{ijk} = 0, \tag{A.10}$$

$$\varepsilon_{imn}\varepsilon_{jmn} = 2\delta_{ij}, \tag{A.11}$$

$$\varepsilon_{ijk}\varepsilon_{mnk} = \delta_{im}\delta_{jn} - \delta_{in}\delta_{jm}. \tag{A.12}$$

The cross product of two vectors $\mathbf{u}$ ($\boldsymbol{u}_i$) and $\mathbf{v}$ ($\boldsymbol{v}_i$),$\mathbf{w} = \mathbf{u} \times \mathbf{v}$, can consequently be written as

$$w_k = \varepsilon_{ijk}u_i v_j. \tag{A.13}$$

A.3 Product Operations of Two Tensors

Given two second-order tensors $\mathbf{A}$ ($\boldsymbol{A}_{ij}$) and $\mathbf{B}$ ($\boldsymbol{B}_{ij}$), we have different types of products: the dyadic product, the single dot product and the double dot product.

The *dyadic product* is a fourth-order tensor, written as

$$C_{ijkl} = A_{ij}B_{kl}, (\mathbf{C} = \mathbf{A} \otimes \mathbf{B}), \tag{A.14}$$

The *single dot product* is a second-order tensor, written as

$$C_{ij} = A_{ik}B_{kj}, (\mathbf{C} = \mathbf{A} \cdot \mathbf{B}), \tag{A.15}$$

The *double dot product* is a scalar, written as

$$C = A_{ij}B_{ji}, (C = \mathbf{A} : \mathbf{B}), \tag{A.16}$$

We define $\mathbf{A}^2 = \mathbf{A}{\cdot}\mathbf{A}$. It can be shown that

$$\mathrm{tr}\mathbf{A}^2 = \mathbf{A} : \mathbf{A} = A_{ij}A_{ji}. \tag{A.17}$$

A.4 Eigenvalue, Eigenvector and Invariants

Give a second-order tensor A_{ij}, if there exists a direction (defined by a non-zero unit vector n_i) and a scalar λ, satisfying

$$A_{ij}n_j = \lambda n_i, \tag{A.18}$$

then the scalar λ is an *eigenvalue* of the tensor, and n_i is called the *eigenvector*. Noting that $n_i = \delta_{ij} n_j$, the above equation can be written as

$$(A_{ij} - \lambda\delta_{ij})n_j = 0, \tag{A.19}$$

which is a system of three linear equations for n_j (i.e., n_1, n_2 and n_3). In order that there be a non-trivial solution for n_j, it is necessary that

$$\det(A_{ij} - \lambda\delta_{ij}) = 0. \tag{A.20}$$

Expansion of the above equation gives the cubic characteristic equation for λ

$$\lambda^3 - I_1\lambda^2 + I_2\lambda - I_3 = 0, \tag{A.21}$$

where

$$I_1 = A_{ii}, \tag{A.22}$$

$$I_2 = \frac{1}{2}\left(A_{ii}A_{jj} - A_{ij}A_{ji}\right), \tag{A.23}$$

$$I_3 = \det A_{ij}, \tag{A.24}$$

where $\det A_{ij}$ stands for the determinant of A_{ij}, which can also be expressed in the form $\varepsilon_{ijk}\, A_{1i}\, A_{2j}\, A_{3k}$. The quantities I_1, I_2 and I_3 are called the first, second, and third invariants of the tensor A_{ij}. Sometimes the following invariants are also used:

$$I = A_{ii}, \tag{A.25}$$

$$II = A_{ij}\, A_{ji}, \tag{A.26}$$

$$III = A_{ij}\, A_{jk}\, A_{ki}. \tag{A.27}$$

A.5 Gradient Operations

A function of the position vector $\mathbf{x}$ is called a field. We can have a scalar field, a vector field or a tensor field. Derivatives with respect to position vectors are performed using the vector differential operator ∇, know as the *del operator*. It is written as $\partial/\partial x_i$ in the Cartesian tensor notation. The operator can be treated as a vector but it cannot stand alone. It must operate on a scalar, vector or a tensor.

Gradient of a scalar The gradient of a scalar field ρ is a vector defined by $\nabla\rho$, or $\partial\rho/\partial x_i$.

Gradient of a vector The gradient of a vector field $\mathbf{u}$ is defined by $\nabla\mathbf{u}$. It should be careful here that the corresponding Cartesian tensor notation is $\partial u_j/\partial x_i$, not $\partial u_i/\partial x_j$ (Our convention assigns subscripts from left to right, and therefore to translate $\nabla\mathbf{u}$ to the Cartesian tensor notation, we write $\partial/\partial x_i$ followed by u_j). The corresponding Gibbs notation for $\partial u_i/\partial x_j$ is $(\nabla\mathbf{u})^{\mathrm{T}}$.

Divergence of a vector The divergence of a vector field $\mathbf{u}$ is a scalar defined by

$$\nabla \cdot \mathbf{u} = \partial u_i / \partial x_i (\equiv \partial u_1 / \partial x_1 + \partial u_2 / \partial x_2 + \partial u_3 / \partial x_3).$$

Curl of a vector The curl of a vector **u** is a vector defined by $\nabla \times \mathbf{u}$, or $\varepsilon_{kij}(\partial u_j/\partial x_i)$.

Divergence of a tensor The divergence of a tensor field **A** is a vector defined by $\nabla \cdot \mathbf{A}$, or $\partial A_{ij}/\partial x_i$.

Appendix B

B.1 Eshelby Tensor Components

When the matrix is isotropic and the inclusion is spheroidal with the symmetric axis identified as x_1, the non-vanishing components of the Eshelby tensor were derived by Tandon and Weng (1984) as

$$E_{1111} = \frac{1}{2(1-\nu^m)}\left[1 - 2\nu^m + \frac{3a_r^2 - 1}{a_r^2 - 1} - \left(1 - 2\nu^m + \frac{3a_r^2}{a_r^2 - 1}\right)g\right], \tag{B.1}$$

$$E_{2222} = \frac{3a_r^2}{8(1-\nu^m)\left(a_r^2 - 1\right)} + \frac{1}{4(1-\nu^m)}\left[1 - 2\nu^m - \frac{9}{4\left(a_r^2 - 1\right)}\right]g, \tag{B.2}$$

$$E_{2233} = \frac{1}{4(1-\nu^m)}\left\{\frac{a_r^2}{2\left(a_r^2 - 1\right)} - \left[1 - 2\nu^m + \frac{3}{4\left(a_r^2 - 1\right)}\right]g\right\}, \tag{B.3}$$

$$E_{2211} = -\frac{a_r^2}{2(1-\nu^m)\left(a_r^2 - 1\right)} + \frac{1}{4(1-\nu^m)}\left[\frac{3a_r^2}{a_r^2 - 1} - (1 - 2\nu^m)\right]g, \tag{B.4}$$

$$E_{1122} = -\frac{1}{2(1-\nu^m)}\left[1 - 2\nu^m + \frac{1}{a_r^2 - 1}\right] + \frac{1}{2(1-\nu^m)}\left[1 - 2\nu^m + \frac{3}{2\left(a_r^2 - 1\right)}\right]g, \tag{B.5}$$

$$E_{2323} = \frac{1}{4(1-\nu^m)}\left\{\frac{a_r^2}{2\left(a_r^2 - 1\right)} + \left[1 - 2\nu^m - \frac{3}{4\left(a_r^2 - 1\right)}\right]g\right\}, \tag{B.6}$$

$$E_{1212} = \frac{1}{4(1-\nu^m)}\left\{1 - 2\nu^m - \frac{a_r^2 + 1}{a_r^2 - 1} - \frac{1}{2}\left[1 - 2\nu^m - \frac{3\left(a_r^2 + 1\right)}{a_r^2 - 1}\right]g\right\}, \tag{B.7}$$

$$E_{3333} = E_{2222},\ E_{3322} = E_{2233},\ E_{3311} = E_{2211},\ E_{1133} = E_{1122},\ E_{1313} = E_{1212}. \tag{B.8}$$

R. Zheng et al., *Injection Molding*, DOI: 10.1007/978-3-642-21263-5,

Table B.1 Relation between indices in contracted and tensor notations

Contracted notation	Tensor notation
1	11
2	22
3	33
4	23
5	13
6	12

where ν^m is Poisson's ratio of the matrix, a_r is the aspect ratio of the fiber and g is given by

$$g = \frac{a_r}{\left(a_r^2 - 1\right)^{3/2}} \left[a_r\left(a_r^2 - 1\right)^{1/2} - \cosh^{-1} a_r\right], \quad \text{for } a_r > 1; \tag{B.9}$$

$$g = \frac{a_r}{\left(1 - a_r^2\right)^{3/2}} \left[\cosh^{-1} a_r - a_r\left(1 - a_r^2\right)^{1/2}\right], \quad \text{for } a_r > 1. \tag{B.10}$$

For a spherical inclusion, $a_r = 1$, one has

$$E_{1111} = E_{2222} = E_{3333} = \frac{7 - 5\nu^m}{15(1 - \nu^m)}, \tag{B.11}$$

$$E_{1122} = E_{2233} = E_{3311} = \frac{5\nu^m - 1}{15(1 - \nu^m)}, \tag{B.12}$$

$$E_{1212} = E_{2323} = E_{3131} = \frac{4 - 5\nu^m}{15(1 - \nu^m)}. \tag{B.13}$$

B.2 Contracted Notation for Stiffness Tensor and Compliance Tensor

Since $\sigma_{ij}$and ε_{ij} are symmetric tensors, each of them has 6 independent components, the fourth order stiffness tensor C_{ijkl} contains at most 36 independent constants, such that it can be displayed as a 6 × 6 matrix of components using contracted notation, C_{mn}, where $m, n = 1, 2, 3,4, 5, 6$. There is a unique correspondence between of C_{mn} and C_{ijkl}. The index m is related to ij, and n is related to kl, as shown in Table B.1. For instance, $C_{1122} = C_{12}$, $C_{1323} = C_{54}$. Since $C_{mn} = C_{nm}$, the number of independent constants is generally 21. For orthotropic materials, the the number of independent constants further reduces to 9. When the fourth order tensor is transformed to the principal axes, all $C_{ij} = 0$, except for C_{11}, C_{22}, C_{33}, C_{12}, C_{13}, C_{23}, C_{44}, C_{55}, and C_{66}.

Similarly, the fourth order elastic compliance tensor S_{ijkl} can be expressed in a contracted form S_{mn}.

B.3 Inverse of a Matrix

In mechanical property calculations of fiber-reinforced polymers, one often needs to calculate the inverse of a matrix P_{ij}.

$$[P_{ij}] = \begin{bmatrix} P_{11} & P_{12} & P_{13} & 0 & 0 & 0 \\ P_{21} & P_{22} & P_{23} & 0 & 0 & 0 \\ P_{31} & P_{32} & P_{33} & 0 & 0 & 0 \\ 0 & 0 & 0 & P_{44} & 0 & 0 \\ 0 & 0 & 0 & 0 & P_{55} & 0 \\ 0 & 0 & 0 & 0 & 0 & P_{66} \end{bmatrix}. \tag{B.14}$$

If there is a matrix Q_{ij}, such that

$$P_{ik}Q_{kj} = \delta_{ij}, \tag{B.15}$$

then the matrix P_{ij} is said to be invertible, and $Q_{ij}$is called the inverse of P_{ij}, written as $Q_{ij} = P_{ij}^{-1}$.

The components of Q_{ij} can be calculated by

$$\begin{aligned} Q_{11} &= (P_{22}P_{33} - P_{23}P_{23})/\det(P_{ij}), \\ Q_{22} &= (P_{11}P_{33} - P_{13}B_{13})/\det(P_{ij}), \\ Q_{33} &= (P_{11}P_{22} - P_{12}P_{12})/\det(P_{ij}), \\ Q_{23} &= (P_{12}P_{13} - P_{11}P_{23})/\det(P_{ij}), \\ Q_{13} &= (P_{12}P_{23} - P_{13}P_{22})/\det(P_{ij}), \\ Q_{12} &= (P_{13}P_{23} - P_{12}P_{33})/\det(P_{ij}), \\ Q_{44} &= 1/P_{44}, \\ Q_{55} &= 1/P_{55}, \\ Q_{66} &= 1/P_{66}. \end{aligned} \tag{B.16}$$

where

$$\begin{aligned} \det(P_{ij}) =& P_{11}(P_{22}P_{33} - P_{23}P_{23}) + P_{12}(P_{13}P_{23} - P_{12}P_{33}) \\ &+ P_{13}(P_{12}P_{23} - P_{13}P_{22}). \end{aligned} \tag{B.17}$$

For example, given the contracted stiffness matrix C_{ij}, one can use the above equations to calculate the contracted compliance matrix S_{ij}, or vice versa,

B.4 Expanded Mori–Tanaka Equation

Here we will consider the expanded expression of the Mori–Tanaka equation, which is useful for practical calculations. It is easy to show (Mori and Wakashima 1989) that Eqs. 6.20 and 6.21 can be rewritten as:

$$C_{ijkl} = C^m_{ijkl} + \varphi^f B^{-1}_{ijkl}, \tag{B.18}$$

with

$$B_{ijkl} = \varphi^m E_{ijmn} S^m_{mnkl} + (C^f_{ijkl} - C^m_{ijkl})^{-1}, \tag{B.19}$$

where E_{ijkl} is the Eshelby tensor, φ^f and φ^m are the volume fractions of fibers and the polymer, respectively, S^m_{ij} is the elastic compliance of the polymer, and C^f_{ij} and C^m_{ij} are the stiffness constants of the fiber and the polymer, respectively.

Using the contracted notation, we write Eq. (B.18) as

$$C_{ij} = C^m_{ij} + \varphi^f B^{-1}_{ij}. \tag{B.20}$$

For orthotropic properties, all $C_{ij} = 0$, except for C_{11}, C_{22}, C_{33}, $C_{23} = C_{32}$, $C_{13} = C_{31}$, $C_{12} = C_{21}$, C_{44}, C_{55}, and C_{66}. From B.19), the components of B_{ij} are given by

$$\begin{aligned}
B_{11} &= \varphi^m \left(E_{1111} S^m_{11} + E_{1122} S^m_{12} + E_{1133} S^m_{13}\right) + \left(C^f_{11} - C^m_{11}\right)^{-1}, \\
B_{22} &= \varphi^m (E_{2211} S^m_{12} + E_{2222} S^m_{22} + E_{2233} S^m_{23}) + (C^f_{22} - C^m_{22})^{-1}, \\
B_{33} &= \varphi^m \left(E_{3311} S^m_{13} + E_{3322} S^m_{23} + E_{3333} S^m_{33}\right) + \left(C^f_{33} - C^m_{33}\right)^{-1}, \\
B_{23} &= \varphi^m \left(E_{2211} S^m_{13} + E_{2222} S^m_{23} + E_{2233} S^m_{33}\right) + \left(C^f_{23} - C^m_{23}\right)^{-1}, \\
B_{13} &= \varphi^m \left(E_{1111} S^m_{13} + E_{1122} S^m_{23} + E_{1133} S^m_{33}\right) + \left(C^f_{13} - C^m_{13}\right)^{-1}, \\
B_{12} &= \varphi^m \left(E_{1111} S^m_{12} + E_{1122} S^m_{22} + E_{1133} S^m_{23}\right) + \left(C^f_{12} - C^m_{12}\right)^{-1}, \\
B_{44} &= 2\varphi^m E_{2323} S^m_{44} + \left(C^f_{44} - C^m_{44}\right)^{-1}, \\
B_{55} &= 2\varphi^m E_{1313} S^m_{55} + \left(C^f_{55} - C^m_{55}\right)^{-1}, \\
B_{66} &= 2\varphi^m E_{1212} S^m_{66} + \left(C^f_{66} - C^m_{66}\right)^{-1}.
\end{aligned} \tag{B.21}$$

B.5 Calculation of Engineering Moduli from Compliance Constants

In dealing with engineering problems, we often desire to convert C_{ijkl} or S_{ijkl} to the engineering moduli (Young's moduli, shear moduli and Poisson's ratios). The engineering moduli are easily calculated from the components of the contracted compliance matrix. The formulas are as follows (There are 9 nonzero independent elastic constants for orthotropic materials):

$$E_{11} = \frac{1}{S_{11}}, E_{22} = \frac{1}{S_{33}}, E_{33} = \frac{1}{S_{33}},$$

$$\upsilon_{12} = -\frac{S_{12}}{S_{11}}, \upsilon_{23} = -\frac{S_{23}}{S_{22}}, \upsilon_{31} = -\frac{S_{13}}{S_{33}},$$

$$G_{12} = \frac{1}{S_{66}}, G_{23} = -\frac{1}{S_{55}}, G_{13} = \frac{1}{S_{44}}, \tag{B.22}$$

For converting C_{ijkl} to engineering moduli, one first converts it to the contracted notation, and takes a matrix inverse to find the corresponding compliance matrix, and then use Eq. B.22 to obtain the engineering constants.

Appendix C

C.1 Gradient Calculation on Constant Elements

In the Hele-Shaw approximation, the in-plane velocity gradient $\partial u/\partial x$ and $\partial u/\partial y$ are negligible compared to $\partial u/\partial z$. However, in some calculations such as the fiber orientation calculation, the information of in-plane velocity gradients is required. In the Hele-Shaw flow simulation, constant elements are commonly used for the velocity, so that the resulting velocity field is constant within an element and discontinuous across the element boundary. A convenient and useful approximation to the gradient within constant element is based on the idea described below. First, the gradient $\partial u/\partial x$ in an element K is taken to be the area average gradient:

$$\left\langle \frac{\partial u}{\partial \mathbf{x}} \right\rangle = \frac{1}{A^{(K)}} \int_{A^{(K)}} \frac{\partial u}{\partial \mathbf{x}} dA, \tag{C.1}$$

where $A^{(K)}$ is the area of element K. The numerical evaluation of Eq. C.1 is facilitated by using the divergence theorem to obtain

$$\left\langle \frac{\partial u}{\partial \mathbf{x}} \right\rangle = \frac{1}{A^{(K)}} \int_{\partial A^{(K)}} u^{(\mathrm{edge})} \mathbf{n} d\Gamma, \tag{C.2}$$

where $\partial A^{(K)}$ is the boundary of the element $A^{(K)}$, $\mathbf{n}$ is the outward normal to $\partial A^{(K)}$at a point in $d\Gamma$, and the value of $u^{(\mathrm{edge})}$on the edge is calculated by taking the average of the values in the two elements on either side of the edge.

For a triangular element in Fig. C.1, we nominate unit vectors $\mathbf{i}$, $\mathbf{j}$, and $\mathbf{k}$ in the Cartesian x, y, and z directions, we have

$$\mathbf{n}_1 = \frac{y_3 - y_2}{l_1}\mathbf{i} + \frac{x_2 - x_3}{l_1}\mathbf{j}, \tag{C.3}$$

$$\mathbf{n}_2 = \frac{y_1 - y_3}{l_2}\mathbf{i} + \frac{x_3 - x_1}{l_2}\mathbf{j}, \tag{C.4}$$

R. Zheng et al., *Injection Molding*, DOI: 10.1007/978-3-642-21263-5,

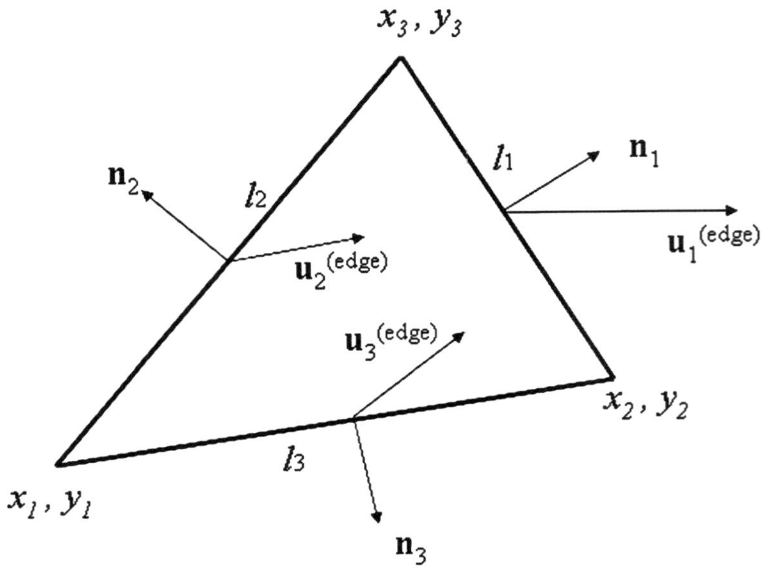

Fig. C.1 Edge velocities used for the velocity gradient calculation in a constant triangular element

$$\mathbf{n}_3 = \frac{y_2 - y_1}{l_3}\mathbf{i} + \frac{x_1 - x_2}{l_3}\mathbf{j}, \tag{C.5}$$

$$\mathbf{u}_1^{(edge)} = u_1^{(edge)}\mathbf{i} + v_1^{(edge)}\mathbf{j}, \tag{C.6}$$

$$\mathbf{u}_2^{(edge)} = u_2^{(edge)}\mathbf{i} + v_2^{(edge)}\mathbf{j}, \tag{C.7}$$

$$\mathbf{u}_3^{(edge)} = u_3^{(edge)}\mathbf{i} + v_3^{(edge)}\mathbf{j}. \tag{C.8}$$

Form Eq. C.2, we obtain

$$\left\langle \frac{\partial u}{\partial x} \right\rangle = \frac{1}{A^{(K)}}\left[(y_3 - y_2)u_1^{(edge)} + (y_1 - y_3)u_2^{(edge)} + (y_2 - y_1)u_3^{(edge)}\right], \tag{C.9}$$

$$\left\langle \frac{\partial u}{\partial y} \right\rangle = \frac{1}{A^{(K)}}\left[(x_2 - x_3)u_1^{(edge)} + (x_3 - x_1)u_2^{(edge)} + (x_1 - x_2)u_3^{(edge)}\right], \tag{C.10}$$

$$\left\langle \frac{\partial v}{\partial x} \right\rangle = \frac{1}{A^{(K)}}\left[(y_3 - y_2)v_1^{(edge)} + (y_1 - y_3)v_2^{(edge)} + (y_2 - y_1)v_3^{(edge)}\right], \tag{C.11}$$

$$\left\langle \frac{\partial v}{\partial y} \right\rangle = \frac{1}{A^{(K)}}\left[(x_2 - x_3)v_1^{(edge)} + (x_3 - x_1)v_2^{(edge)} + (x_1 - x_2)v_3^{(edge)}\right], \tag{C.12}$$

with

$$
\begin{aligned}
A^{(K)} &= \frac{1}{2}\mathbf{k} \cdot \det \begin{bmatrix} \mathbf{i} & \mathbf{j} & \mathbf{k} \\ x_2 - x_1 & y_2 - y_1 & 0 \\ x_3 - x_1 & y_3 - y_1 & 0 \end{bmatrix} \\
&= \frac{1}{2}[(x_2 - x_1)(y_3 - y_1) - (x_3 - x_1)(y_2 - y_1)].
\end{aligned}
\tag{C.13}
$$

Specially, if an edge of a triangular element coincides with the edge of the cavity, where no neighboring element is sharing this edge, a zero $\mathbf{u}^{(\text{edge})}$ is specified on this edge to impose the no-slip boundary condition.

The above described gradient approximation method sometimes can introduce undue oscillations. One may use monotonicity algorithms (see, for example, van Leer 1979) to preserve monotonicity.

References

Advani SG, Tucker CL III (1987) The use of tensors to describe and predict fiber orientation in short fiber composites. J Rheol 31:751–784

Advani SG, Tucker CL III (1990) Closure approximations for three-dimensional structure tensors. J Rheol 34:367–386

Agassant JF, Avenas P, J-Ph Sergent, Carreau PJ (1991) polymer processing: Principles and modeling. Hanser, Munich

Ahmed S, Irons BM, Zienkiewicz OC (1970) Analysis of thick and thin shell structures by curved finite elements. Int J Numer Meth Eng 2:419–451

Akczurowski E, Mason SG (1968) Particle motions in sheared suspensions XXIV rotation of spheroids and cylinders. Trans Soc Rheol 12:209–215

Albert C, Fernlund G (2002) Spring-in and warpage of angled composite laminates. Compos Sci Technol 62:1895–1912

Al-Hussein M, Strobl G (2002) The melting line, the crystallization line and the equilibrium melting temperature of isotactic polystyrene. Macromol 35:1672–1676

Allen MP, Tildesley DJ (1989) Computer simulation of liquids. Oxford University Press, New York

Ammar A, Leo V, Régnier G (2003) Corner deformation of injected thermoplastic parts. Int J Forming Processes 6:53–70

Antosiewitcz HA (1972) Bessel functions of fractional order. In: Abramowitz M, Stegun IA (eds) Handbook of mathematical functions with formulas, graphs, and mathematical tables, 9th printing. Dover, New York, pp 437–478

Arnold DN, Brezzi F, Fortin M (1984) A stable finite element for the stokes equations. Calcolo 21:337–344

Austin C (1991) Warpage design principles. Moldflow Pty Ltd, Melbourne

Austin C (1987) Computer aided engineering in injection molding. In: Manzione LT (ed) Applications of computer aided engineering in injection molding. Hanser, New York, pp 137–172

Avrami M (1939) Kinetics of phase change, I. General theory. J Chem Phys 7:1103–1112

Baaijens FPT (1991) Calculation of residual stress in injection molded products. Rheol Acta 30:284–299

Bakharev A, Fan Z, Costa FS (2004) Prediction of core shift effects using mold filling simulation. Society of Plastics Engineers Annual Technical Conference—ANTEC' 2004(50):621–625

Bakharev A, Zheng R, Fan Z, Costa FS, Jin X, Kennedy PK (2005) Corner effect in warpage simulation. Society of Plastics Engineers Annual Technical Conference—ANTEC' 2005(51):511–515

R. Zheng et al., *Injection Molding*, DOI: 10.1007/978-3-642-21263-5,

Baltussen MGHM, Hulsen MA, Peters GWM (2010) Numerical simulation of the fountain flow instability in injection molding. J Non-Newtonian Fluid Mech 165:631–640
Batoz JL, Lardeur P (1989) A discrete shear triangular nine d.o.f. element for the analysis of thick to very thin plates. Int J Numer Meth Eng 28:533–560
Bay RS (1991) Orientation in injection molded composites: A comparison of theory and experiment. Ph.D Thesis, University of Illinois, Urbana, II
Benveniste Y (1987) A new approach to the application of Mori-Tanaka's theory in composite materials. Mech Mater 6:147–157
Bergan PG, Felippa CA (1985) A triangular membrane element with rotational degrees of freedom. Comput Meth in Appl Mech Eng 50:25–69
Berger JL, Gogos CG (1973) A Numerical simulation of the cavity filling process with PVC in injection molding. Polym Eng Sci 13:102–112
Bird RB, Armstrong RC, Hassager O (1987a) Dynamics of polymeric liquids, Volume 1: Fluid Mechanics. John Wiley and Sons, New York
Bird RB, Curtiss CF, Armstrong RC, Hassager O (1987b) Dynamics of polymeric liquids, Volume 2: Kinetic Theory. Volume 2: John Wiley and Sons, New York
Bogaerds ACB, Hulsen MA, Peters GWM, Baaijens FPT (2004) Stability analysis of injection molding flows. J Rheol 48:765–785
Boger DV, Walters K (1993) Rheological phenomena in focus. Elsevier, Amsterdam
Boitout F, Agassant JF, Vincent M (1995) Elastic calculation of residual stresses in injection molding. Int Polym Process 10:237–242
Boutaous M, Bourgin P, Zinet M (2010) Thermally and flow induced crystallization of polymers at low shear rate. J Non-Newtonian Fluid Mech 165:227–237
Brebbia CA, Nardini D (1986) Solution of parabolic and hyperbolic time dependent problems using boundary elements. Comp Maths with Appls 12B:1061–1072
Brebbia CA, Walker S (1980) Boundary element techniques in engineering. Newnes-Butterworth, London
Brincat P, Friedl C, Talwar K (1998) Extensional viscosity modeling for injection molding simulation. Society of Plastics Engineers Annual Technical Conference—ANTEC'1998 44:552–556
Bush MB, Tanner RI (1990) Boundary element analysis of slow non-Newtonian flow. In: Banerjee PK, Morino L (eds) Boundary element methods in nonlinear fluid dynamics. Elsevier Applied Science, London, pp 285–317
Bushko WC, Stokes VK (1995) Solidification of thermoviscoelastic melts: part 1. formulation of model problem. Polym Eng Sci 35:351–364
Buzbee BL, Dorr FW, George JA, Golub GH (1971) The direct solution of the discrete poisson equation on irregular regions. SIAM J Numer Anal 8:722–738
Caspers L (1996) VIp, An integral approach to the simulation of injection molding. Ph.D Thesis, Eindhoven University of Technology
Chang RY, Yang WH (2001) Numerical simulation of mold filling in injection molding using a three-dimensional finite volume approach. Int J Numer Meth Fluids 37:125–148
Chen SC, Chung YC (1994) Numerical simulations of the cyclic, transient mold Heat transfer in injection mold cooling process. Int Commun Heat and Mass Trans 21:323–332
Chiang HH, Hieber CA, Wang KK (1991) A unified simulation of the filling and postfilling stages in injection molding, Part I: Formulation. Polym Eng Sci 31:116–124
Christensen RM (1990) A critical evaluation for a class of micromechanics models. J Mech Phys Solids 38:379–404
Chung DH, Kwon TH (2002) Invariant-based optimal fitting closure approximation for the numerical prediction of flow-induced fiber orientation. J Rheol 46:169–194
Chung ST, Kwon TH (1995) Numerical simulation of fiber orientation in injection molding of short-fiber-reinforced thermoplastics. Polym Eng Sci 35:604–618
Chung ST, Kwon TH (1996) Coupled analysis of injection molding filling and fiber orientation, including in-plane velocity gradient effect. Polym Compos 17:859–872

Cintra JS, Tucker CL III (1995) Orthotropic closure approximations for flow-induced fiber orientation. J Rheol 39:1095–1121

Cleary P, Ha J, Alguine V, Nguyen T (2002) Flow modeling in casting processes. Appl Math Model 26:171–190

Cleary P, Monaghan JJ (1999) Conduction modelling using smoothed particle hydrodynamics. J Comput Phys 148:227–264

Coates PD (ed) (2007) Polymer process engineering 07: Enhanced polymer processing. University of Bradford, Bradford

Coleman CJ (1990) A boundary element approach to some non-linear equations from fluid mechanics. Comput Mech 6:197–202

Coleman CJ, Tullock D, Phan-Thien N (1991) An effective boundary element method for inhomogeneous partial differential equations. Z Angew Math Phys 42:730–745

Coppola S, Grizzuiti N, Maffettone PL (2001) Microrheological modeling of flow-induced crystallization. Macromol 34:5030–5036

Coyle DJ, Blake JW, Macosko CW (1987) The kinematics of fountain flow in mold filling. AIChE J 33:1168–1177

Crochet MJ, Davies AR, Walters K (1984) Numerical simulation of non-newtonian flow. Elsevier, Amsterdam

Crochet MJ, Dupret F, Verleye V (1994) Injection molding. In: Advani SG (ed) Flow and rheology in polymer composites manufacturing. Elsevier, Amsterdam, pp 442–444

Cross MM (1965) Rheology of non-Newtonian fluids: a new flow equation for pseudoplastic systems. J Colloid Sci 20:417–437

Curran DAS, Cross MM, Lewis BA (1980) Solution of parabolic differential equations by the boundary element method using discretization in time. Appl Math Model 4:398–400

Dai SC, Qi F, Tanner RI (2006) Strain and strain-rate formulation for flow-induced crystallization. Polym Eng Sci 46:659–669

Dai SC (2006) Tanner RI Anisotropic thermal conductivity in sheared polypropylene. Rheol Acta 45:228–238

De Gennes PG (1979) Scaling concepts in polymer physics. Cornell University Press, Ithaca

Dealy JM (1994) Official nomenclature for material functions describing the response of a viscoelastic fluid to various shearing and extensional deformations. J Rheol 38:179–191

Debbaut B, Crochet MJ (1988) Extensional effects in complex flows. J Non-Newtonian Fluid Mech 30:169–184

Delaunay D, Le Bot PH, Fulchiron R, Luyé JF, Regnier G (2000a) Nature of contact between polymer and mold in injection molding, Part I: influence of a non-perfect thermal contact. Polym Eng Sci 40:1682–1691

Delaunay D, Le Bot PH, Fulchiron R, Luyé JF, Regnier G (2000b) Nature of contact between polymer and mold in injection molding. Part II: Influence of mold deflection on pressure history and shrinkage. Polym Eng Sci 40:1692–1700

Denn MM (2001) Extrusion instabilities and wall slip. Annu Rev Fluid Mech 33:265–287

Denn MM (2004) Fifty years of non-Newtonian fluid dynamics. AIChE J 50:2335–2345

Dinh SH, Armstrong RC (1984) A rheological equation of state for semi-concentrated fiber suspensions. J Rheol 28:207–227

Doi M (1981) Molecular dynamics and rheological properties of concentrated solution of rodlike polymers in isotropic and liquid crystalline phases. J Polym Sci: Polym Phys Edition 19: 229-143

Doi M, Edwards SF (1986) The theory of polymer dynamics. Clarendon Press, Oxford

Donnell LH (1976) Beams, plates and shells. McGraw Hill, New York

Dou HS, Phan-Thien N, Khoo BC, Yeo KS, Zheng R (2004) Simulation of front evolving liquid film flowing down an inclined plate using level set method. Comput Mech 34:271–281

Dou HS, Khoo BC, Phan-Thien N, Yeo KS, Zheng R (2007) Simulations of fiber orientation in dilute suspensions with front moving in the filling process of a rectangular channel using level set method. Rheol Acta 46:427–447

Doufas AK, Dairanieh IS, McHugh AJ (1999) A continuum model for flow-induced crystallization of polymer melts. J Rheol 43:85–109

Doufas AK, McHugh AJ, Miller C (2000) Simulation of melt spinning including flow-induced crystallization, Part I. Model development an predictions. J Non-Newtonian Fluid Mech 92:27–66

Dowson D (1962) A generalized Reynolds equation for fluid film lubrication. Int J Mech Sci 4:159–170

Dupret F, Vanderschuren L (1988) Calculation of the temperature field in injection molding. AICHE J 34:1959–1972

Dupret F, Verleye V (1999) Modelling the flow of fiber suspensions in narrow gaps. In: Siginer DA, De Kee D, Chhabra RP (eds) Advances in the flow and rheology of Non-Newtonian fluids, Part 2, vol 8. Elsevier, Amsterdam, pp 1347–1398

Dutta NK, Edward G (1997) Generic relaxation spectra of solid polymers: 1. Development of spectral distribution model and its application to stress relaxation of polypropylene. J Appl Polym Sci 66:1101–1115

Eder G, Janeschitz-Kriegl H (1997) Crystallization. In: Meijer HEH (ed) Materials science and technology, processing of polymers, vol 18. Wiley-VCH, New York, pp 189–268

Eduljee RF, McCullough RL, Gillespie JW Jr (1994) The influence of aggregated and dispersed textures on the elastic properties of discontinuous-fiber composites. Compos Sci Technol 50:381–391

Einstein A (1906) A new determination of molecular dimensions. Annu Physik 19:289–306. Corrections, ibid. (1911) 34:591-592. In: Fürth R (1956) (ed) Investigations on the Theory of the Brownian Movement (Translated by Cowper AD). Dover, New York

Ellero M, Tanner RI (2005) SPH simulations of transient viscoelastic flows at low Reynolds number. J Non-Newtonian Fluid Mech 132:61–72

Ericksen JL (1960) Anistropic fluids. Arch Rat Mech Anal 4:231–237

Eshelby JD (1957) The determination of the elastic field of an ellipsoidal inclusion, and related problems. Proc Roy Soc A 241:376–396

Evans DJ, Holian BL (1985) The Nose-Hoover thermostat. J Chem Phys 83:4069–4074

Fan XJ (2006) Numerical study on some rheological problems of fiber suspensions. Ph.D Thesis, Sydney University. See also, Fan XJ; Phan-Thien N, Tanner RI (2008) Numerical study on some rheological problems of fiber suspensions, VDM Verlag, Saarbrücken, Germany

Fan XJ, Phan-Thien N, Zheng R (1998) A direct simulation of fiber suspensions. J Non-Newtonian Fluid Mech 74:113–136

Fan XJ, Phan-Thien N, Zheng R (1999) Simulation of fiber suspension flows by the Brownian field method. J Non-Newtonian Fluid Mech 84:257–274

Fan XJ, Phan-Thien N, Zheng R (2000a) Simulation of fiber suspension flow with shear-induced migration. J Non-Newtonian Fluid Mech 90:47–63

Fan XJ, Tanner RI, Zheng R (2010a) Smoothed particle hydrodynamics simulation of non-Newtonian molding flow. J Non-Newtonian Fluid Mech 165:219–226

Fan XJ, Tanner RI, Zheng R (2011) Smoothed particle hydrodynamics and its application to non-newtonian molding flow. In: Kauffer PH (ed) Injection molding: process, design, and applications. Nova Science Publishers, New York, pp 101–153

Fan XJ, Zheng R, Liu GW, Tanner RI, Edward G (2010b) Modelling post-molding warping: Part 1: Volumetric expansion. Int Polym Process 25:41–46

Fan XJ, Zheng R, Liu GW, Tanner RI, Edward G (2010c) Modelling post-molding warping: Part 2: Deformation. Int Polym Process 25:47–54

Fan Y, Dai S, Tanner RI (2003) Rheological properties of some thermotropic liquid crystalline polymers. Korea-Aust Rheol J 15:109–115

Fan Z, Friedl C, Costa FS, Jin X (2010d) Three-dimensional anisotropic thermoviscoelastic simulation of residual stresses in injection moldings of thermoplastics. Proc 9th World Congr Comput Mech, Sydney (Abstract, p 69)

Fan Z, Iwamura C, Xu S, Costa FS, Supramaniam Y (2006) Reducing the computation time for three-dimensional simulation of warpage of injection molded plastic parts. Proc 7th World Congr. Comput. Mech. Los Angeles, CA (Abstract p 343)

Fan Z, Lin B, Costa FS, Jin X, Zheng R, Kennedy PK (2004a) Three-dimensional warpage simulation for injection molding. Society of Plastics Engineers Annual Technical Conference—ANTEC' 2004(50):493–495

Fan Z, Kietzmann C, Ray SR, Costa FS, Kennedy PK (2005) Three-dimensional cooling and warpage simulation for the Injection over-molding process. Society of Plastics Engineers Annual Technical Conference—ANTEC' 2005(51):568–572

Fan Z, Zheng R, Kennedy PK, Yu H, Bakharev A (2000b) Warpage analysis of solid geometry. Society of Plastics Engineers Annual Technical Conference—ANTEC' 2000(46):723–726

Fan Z, Zheng R, Yu H, Kennedy PK (2004b) Apparatus and method for structural analysis. US Patent No. 6704693

Fang JN, Owens RG, Tacher L, Parriaux A (2006) A numerical study of the SPH method for simulating transient viscoelastic free surface flow. J Non-Newtonian Fluid Mech 139:68–84

Feng J, Leal LG (1997) Simulating complex flows of liquid crystalline polymer using the Doi theory. J Rheol 41:1317–1335

Férec J, Ausias G, Heuzey MC, Carreau PJ (2009) Modeling fiber interactions in semiconcentrated fiber suspensions. J Rheol 53:49–72

Ferrari A, Dumbser M, Toro EF, Armanini A (2009) A new 3D parallel SPH scheme for free surface flows. Comput Fluids 38:1203–1217

Ferry JD (1980) Viscoelastic properties of polymers. Wiley, New York

Folgar FP, Tucker CL III (1984) Orientation behavior of fiber in concentrated suspensions. J Reinforced Plastic Compos 3:98–119

Forstner R, Peters GWM, Rendina C, Housmans JW, Meijer HEH (2009) Volumetric rheology of polymers: The influence of shear flow, cooling rate, and pressure on the specific volume of iPP and P/E random copolymers. J Therm Anal Calorim 98:683–691

Fox TG, Flory PJ (1954) The glass temperature and related properties of polystyrene: influence of molecular weight. J Polym Sci 14:315–319

Friedl C, Costa FS, Cook P, Talwar K, Antanovski LK (2004) Method and apparatus for modeling injection of a fluid in a mold cavity. US Patent, US 6816820 B1

Fulchiron R, Koscher E, Poutot G, Delaunay D, Regnier G (2001) Analysis of the pressure effect on the crystallization kinetics, dilatometric measurements and thermal gradient modeling. J Macromol Sci—Physics 40:297–314

Gadala-Maria F, Acrivos A (1980) Shear-induced structure in a concentrated suspension of solids spheres. J Rheol 24:799–811

Gavazzi AC, Lagoudas DC (1990) On the numerical evaluation of Eshelby's tensor and its application to elastoplastic fibrous composites. Comput Mech 7:13–19

Ghoneim H, Hieber CA (1997) Incorporation of density relaxation in the analysis of residual stresses in molded parts. Polym Eng Sci 37:219–227

Giesekus H (1982) A simple constitutive equation for polymer fluids based on the concept of deformation-dependent tensorial mobility. J Non-Newt Fluid Mech 11:69–109

Giesekus H (1983) Stressing behavior in simple shear flow as predicted by a new constitutive model for polymer fluids. J Non-Newt Fluid Mech 12:367–374

Gingold RA, Monaghan JJ (1977) Smoothed particle hydrodynamics: theory and application to non-spherical stars. Mon Not R Astron Soc 181:375–389

Gordon RJ, Schowalter WR (1972) Anisotropic fluid theory: a different approach to the dumbbell theory of dilute polymer solutions. Trans Soc Rheol 16:79–97

Greco F, D'Avino G, Maffettone PL (2005) Stress tensor of a dilute suspension of spheres in a viscoelastic liquid. Phys Rev Lett 95: Art. No. 24600. Erratum in: Phys Rev Lett (2007) Art. No. 109904

Gruau C, Coupez T (2005) 3D tetrahedral, unstructured and anisotropic mesh generation with adaptation to natural and multidomain metric. Comput Methods Appl Mech Eng 194:4951–4976

Gupta M (2000) Effect of elongational viscosity on axisymmetric entrance flow of polymers. Polym Eng Sci 40:23–35

Hadinata C, Gabriel C, Ruellman M, Laun HM (2005) Comparison of shear-induced crystallization behavior of PB-1 samples with different molecular weight distribution. J Rheol 49:327–349

Hadinata C, Zheng R, Kennedy PK, Zhu P, Edward G, Lee Wo D, Tanner RI (2008) The effects of additives on material properties and shrinkage characteristics. Keynote Lecture at 24th Annual Conference of the Polymer Processing Society, Salerno

Hagerman EM (1973) Weld-line fracture in molded parts. Plast Eng 29:67–69

Hand GL (1962) A theory of anisotropic fluids. J Fluid Mech 13:33–46

Harris JB, Pittma TFT (1975) Equivalent ellipsoidal axis ratios of slender rod-like particles. J Coll Interf Sci 50:280–282

Harry DH, Parrott RG (1970) Numerical simulation of injection mold filling. Polym Eng Sci 10:209–214

Hatzikiriakos SG (1993) A slip model for linear polymers based on adhesive failure. Int Polym Process 8:135–142

Hatzikiriakos SG, Dealy JM (1991) Wall slip of molten high density polyethylene. I. Sliding plate rheometer studies. J Rheol 35:497–523

Hatzikiriakos SG, Dealy JM (1992) Wall slip of molten high density polyethylene. II. Capillary rheometer studies. J Rheol 36:703–741

Henry DP Jr, Banerjee PK (1988) A new BEM formulation for two- and three-dimensional elastoplasticity using particular integrals. Int J Numer Methods Eng 26:2079–2098

Hétu JF, Gao DM, Garcia-Rejon A, Salloum G (1998) 3D finite element method for the simulation of the filling stage in injection molding. Polym Eng Sci 38:223–236

Hétu JF, Lauzé Y, Garcia-Rejon A (1995) Three-dimensional finite element simulation of mold filling processes. In: Shen SF, Dawson PR (eds) Simulation of materials processing: theory, methods and applications, NUMIFORM 95. Balkema, Rotterdam, pp 1135–1140

Hieber CA (2002) Modeling/simulating the injection molding of isotactic polypropylene. Polym Eng Sci 42:1387–1409

Hieber CA, Chiang HH (1992) Shear-rate-dependence modeling of polymer melt viscosity. Polym Eng Sci 32:931–938

Hieber CA, Shen SF (1980) A finite-element/finite-difference simulation of the injection-molding filling process. J Non-Newtonian Fluid Mech 7:1–32

Hill R (1963) Elastic properties of reinforced solids: Some theoretical principles. J Mech Phys Solids 11:357–372

Himasekhar K, Lottey J, Wang KK (1992) CAE of mold cooling in injection molding using a three-dimensional numerical simulation. J Eng Ind Trans ASME 144:213–221

Hinch EJ, Leal LG (1976) Constitutive equations in suspension mechanics, Part 2, Approximate forms for a suspension of rigid particles affected by Brownian rotations. J Fluid Mech 76:187–208

Hirt CW, Nichols BD (1981) Volume of fluid (VOF) method for the dynamics of free boundaries. J Comput Phys 39:201–225

Hobbs SY (1996) The development of flow instabilities during the injection molding of multicomponent resins. Polym Eng Sci 36:1489–1494

Hoffman JD, Weeks JJ (1962) Melting process and the equilibrium melting temperature of polychlorotrifluoroethylene. J Res Natl Bur Stand 66 A:13–28

Holman JP (1976) Heat transfer, 4th edn. McGraw-Hill, New York

Hoover WG (1985) Canonical dynamics: equilibrium phase-space distributions. Phys Rev A31:1695–1697

Hosseini SM, Manzari MT, Hannani SK (2007) A fully explicit three-step SPH algorithm for simulation of non-Newtonian fluid flow. Int J Numer Method Heat and Fluid Flow 17:715–735

Housiadas KD, Tanner RI (2009) On the rheology of a dilute suspension of rigid spheres in a weakly viscoelastic matrix fluid. J Non-Newtonian Fluid Mech 2009 162:88–92

Hughes TJR, Liu WK, Zimmermann TK (1981) Lagrangian-Eulerian finite element formulation for incompressible viscous flows. Comput Methods Appl Mech Eng 29:329–349

Huilgol RR (2006) On the derivation of the symmetric and asymmetric Hele–Shaw flow equations for viscous and viscoplastic fluids using the viscometric fluidity function. J Non-Newtonian Fluid Mech 138:209–213

Huilgol RR, Phan-Thien N (1997) Fluid mechanics of viscoelasticity. Elsevier, Amsterdam

Huilgol RR, Phan-Thien N, Zheng R (1992) A theoretical and numerical study of non-Fourier effects in viscometric and extensional flows of an incompressible simple fluid. J Non-Newtonian Fluid Mech 43:83–102

Huilgol RR, You Z (2006) On the importance of the pressure dependence of viscosity in steady non-isothermal shearing flows of compressible and incompressible fluids and in the isothermal fountain flow. J Non-Newtonian Fluid Mech 136:106–117

Hulsen MA, Van Heel APG, Van den Brule BHAA (1997) Simulation of viscoelastic flows using Brownian configuration fields. J Non-Newtonian Fluid Mech 70:79–101

Ingber MS, Mondy LA (1994) A numerical study of three-dimensional Jeffery orbits in shear flow. J Rheol 38:1829–1843

Ingber MS, Phan-Thien N (1992) A boundary element approach for parabolic differential equations using a class of particular solutions. Appl Math Model 16:124–132

Isayev AI (1987) Orientation, residual stresses, and volumetric effect in injection molding. In: Isayev AI (ed) Injection compression molding fundamentals. Marcel Dekker Inc New York, pp 227-328

Jabbarzadeh A, Atkinson JD, Tanner RI (1999) Wall slip in the molecular dynamics simulation of thin films of hexadecane. J Chem Phys 110:2612–2620

Jabbarzadeh A, Atkinson JD, Tanner RI (2000) The effect of the wall roughness on slip and rheological properties of hexadecane in molecular dynamics simulation of Couette shear flow between two sinusoidal walls. Phys Rev E 61:690–699

Jabbarzadeh A, Tanner RI (2006) Molecular dynamics simulation and its application in nano-rheology. In: Binding DM, Walters K (eds) Rheology Reviews. British Society of Rheology, Aberystwyth, Wales, pp 165–216

Jabbarzadeh A, Tanner RI (2009) Crystallization of alkanes under quiescent and shearing conditions. J Non-Newtonian Fluid Mech 160:11–21

Jabbarzadeh A, Tanner RI (2010) Flow-induced crystallization: unraveling the effects of shear rate and strain. Macromol 43:8136–8142

Janeschitz-Kriegl H (1977) Injection molding of plastics: some ideas about the relationship between mould filling and birefringence. Rheol Acta 16:327–339

Janeschitz-Kriegl H (1979) Injection molding of plastics II: analytical solution of heat transfer problem. Rheol Acta 18:693–701

Janeschitz-Kriegl H (2009) Crystallization Modalities in Polymer Melt Processing. Springer, New York

Janeschitz-Kriegl H, Ratajski E, Stadlbauer M (2003) Flow as an effective promotor of nucleation in polymer melts: a quantitative evaluation. Rheol Acta 42:355–364

Jansen KMB (1994) Residual stresses in quenched and injection molding products. Int Polym Process 9:82–89

Jeffery GB (1922) The motion of ellipsoidal particles immersed in viscous fluid. Proc Roy Soc Lond A102:161–179

Jin X (1993) Boundary element study on particle orientation caused by the fountain flow in injection molding. Polym Eng Sci 33:1238–1242

Johannaber F (1994) Injection molding machines: A user's guidee, 3rd edn. Hanser Publishers, New York

Joseph DD, Liu YJ (1993) Orientation of long bodies falling in a viscoelastic liquid. J Rheol 37:961–981

Kabanemi KK, Crochet MJ (1992) Thermoviscoelastic calculation of residual stresses and residual shapes of injection molded parts. Int Polym Process 7:60–70

Kaelble DH (1985) Computer-aided design of polymers and composites. Marcel Dekker, New York

Kalika DS, Denn MM (1987) Wall slip and extrudate distortion in linear low-density polyethylene. J Rheol 31:815–834

Kamal MR (1987) The McGill University model (McKam-II): a comprehensive, integrated computer simulation of the injection molding process. In: Manzione LT (ed) Application of Computer Aided Engineering in Injection Molding. Hanser, New York

Kamal MR, Chu E, Lafleur PG, Ryan ME (1986) Computer simulation of injection mold filling for viscoelastic melts with fountain flow. Polym Eng Sci 26:190–196

Kamal MR, Goyal SK, Chu E (1988) Simulation of injection mold filling of viscoelastic polymer with fountain flow. AIChE J 34:94–106

Kamal MR, Isayev AI, Liu SJ (eds) (2009) Injection molding—technology and fundamentals. Hanser, Munich

Kamal MR, Kenig S (1972) The injection molding of thermoplastics part I: theoretical model. Polym Eng Sci 12:294–301

Kamal MR, Lai-Fook RA, Hernandez-Aguilar JR (2002) Residual thermal stresses in injection moldings of thermoplastics: a theoretical and experimental study. Polym Eng Sci 42:1098–1114

Kennedy PK (1995) Flow analysis of injection molds. Hanser Publishers, New York

Kennedy PK (2008) Practical and Scientific Aspects of Injection Molding Simulation. Ph.D thesis, Techniche Universiteit Eindhoven

Kennedy PK, Yu H (1997) Plastic-CAE analysis of solid geometry. Society of Plastics Engineers Annual Technical Conference—ANTEC' 97(43):666–669

Kennedy PK, Zheng R (2002) High accuracy shrinkage and warpage prediction for injection molding. Society of Plastics Engineers Annual Technical Conference—ANTEC' 2002, San Francisco, 48:1-7

Kennedy PK, Zheng R (2006) Shrinkage of injection molded material. In: Greener J, Wimberger-Friedl R (eds) Precision injection molding: process, materials, and applications. Hanser, Munich

Keunings R (1986) On the High Weissenberg number problem. J Non-Newtonian Fluid Mech 20:209–226

Keunings R (1997) On the Peterlin approximation for finitely extensible dumbbells. J Non-Newtonian Fluid Mech 68:85–100

Kietzmann C, Van der Walt JP, Morsi YS (1998) A free-front tracking algorithm for a control-volume-based Hele-Shaw method. Int J Numer Methods Eng 41:253–269

Kim SH, Kim CH, Oh H, Choi CH, Kim BY, Youn JR (2007) Residual stresses and viscoelastic deformation of an injection molded automotive part. Korea-Australia Rheol J 19:183–190

Klein DH, Leal LG, García-Cervera CJ, Ceniceros HD (2008) Three-dimensional shear-driven dynamics of polydomain textures and disclination loops in liquid crystalline polymers. J Rheol 52:837–863

Koh CJ, Leal LG, Hookham PA (1994) An experimental investigation of concentrated suspension flow in a rectangular channel. J Fluid Mech 256:1–32

Kolmogoroff AN (1937) On a statistical theory of crystallization of melts, Bull, Akad Sci USSR. Class Sci Math Nat 1:355–359

Koscher E, Fulchiron R (2002) Influence of shear on polypropylene crystallization: morphology development and kinetics. Polym 43:6931–6942

Larson RG (1992) Instabilities in viscoelastic flows. Rheol Acta 31:213–263

Larson RG (1999) The structure and rheology of complex fluids. Oxford University Press, New York

Lauritzen SI, Hoffman JD (1960) Theory of formation of polymer crystals with folded chains in dilute solution. J Res Natl Bur Stand 64A:73–102

Le Meins J-F, Moldenaers P, Mewis J (2003) Suspensions of monodisperse spheres in polymer melts: particle size effects in extensional flow. Rheol Acta 42:184–190

Lee BJ, Mear ME (1991) Effect of inclusion shape on the stiffness of nonlinear two-phase composites. J Mech Phys Solids 39:627–649

Lee Wo D, Tanner RI (2010) The impact of blue organic and inorganic pigments on the crystallization and rheological properties of isotactic polypropylene. Rheol Acta 49:75–88

Lees AW, Edwards SF (1972) The computer study of transport processes under extreme conditions. J Phys C: Solid State Physics 5:1921–1929

Leighton D, Acrivos A (1987a) Measurement of shear-induced self-diffusion in concentrated suspensions of spheres. J Fluid Mech 177:109–131

Leighton D, Acrivos A (1987b) The shear-induced migration of particles in concentrated suspensions. J Fluid Mech 181:415–439

Leo V, Cuvelliez CH (1996) The effect of the packing parameters, gate geometry, and mold elasticity on the final dimensions of a molded part. Polym Eng Sci 36:1961–1971

Liedauer S, Eder G, Janeschitz-Kriegl H, Jerschow P, Geymayer W, Inglic E (1993) On the kinetics of shear induced crystallization in polypropylene. Int Polym Process 8:236–244

Lim FJ, Schowalter WR (1989) Wall slip of narrow molecular weight distribution polybutadienes. J Rheol 33:1359–1382

Lin B, Jin X, Zheng R, Costa FS, Fan Z (2004) 3D fiber orientation simulation for plastic injection molding. NUMIFORM 2004. AIP Conference Proceedings,vol 712, pp 282–287

Lipscomb DD II, Denn MM, Hur DU, Boger DV (1988) The flow of fiber suspension in complex geometry. J Non-Newtonian Fluid Mech 26:297–325

Liu GR, Liu MB (2003) Smoothed particle hydrodynamics, a meshfree particle method. World Scientific, Singapore

Liu MB, Liu GR (2010) Smoothed particle hydrodynamics (SPH): An overview and recent developments. Arch Comput Methods Eng 17:25–76

Liu YJ (2009) Fast multipole boundary element method—theory and applications in engineering. Cambridge University Press, Cambridge

Lodge AS (1964) Elastic Liquids. Academic Press, London

Lord HA, Williams G (1975) Mold-filling studies for the injection molding of thermoplastic materials. Part II: the transient flow of plastic materials in the cavities of injection-molding dies. Polym Eng Sci 15:569–582

Lucy LB (1977) A Numerical approach to testing of fission hypothesis. Astron J 82:1013–1024

Luyé JF, Regnier G, Le Bot PH, Delaunay D, Fulchiron R (2001) PVT Measurement Methodology for semicrystalline polymers to simulate injection-molding process. J Appl Polym Sci 79:302–311

Mackley MR, Wannaborworn S, Gao P, Zhao F (1999) The optical microscopy of sheared liquids using a newly developed optical stage. J Microsc Anal 69:25–27

Mall-Gleissle SE, Gleissle W, McKinley GH, Buggisch H (2002) The normal stress behavior of suspensions with viscoelastic matrix fluids. Rheol Acta 41:61–76

Marand H, Xu J, Srinivas S (1998) Determination of the equilibrium melting temperature of polymer crystals: linear and non-linear Hoffman-Weeks extrapolation. Macromol 31:8219–8229

Mavridis H, Hrymak AN, Vlachopoulos J (1986) Finite element simulation of fountain flow in injection molding. Polym Eng Sci 26:449–454

Mavridis H, Hrymak AN, Vlachopoulos J (1988) The effect of fountain flow on molecular orientation in injection molding. J Rheol 32:639–663

McLeish TCB, Larson RG (1998) Molecular constitutive equations for a class of branched polymers: the Pom–Pom polymer. J Rheol 42:81–110

Mendoza R, Régnier G, Seiler W, Lebrun JL (2003) Spatial distribution of molecular orientation in injection molded iPP: influence of processing conditions. Polym 44:3363–3373
Mitsoulis E (2010) Fountain flow of pseudoplastic and viscoplastic fluids. J Non-Newtonian Fluid Mech 165:45–55
Mlekusch B (2001) Calculation of residual stress development in injection molding using a nonlinear viscoelastic model. Mech. Time-Depend. Mater 5:101–118
Monaghan JJ (1992) Smoothed particle hydrodynamics. Annu Rev Astron Astrophys 30:543–574
Monaghan JJ (1994) Simulating free surface flows with SPH. J Comput Phys 110:399–406
Monaghan JJ (2000) SPH without a tensile instability. J Comput Phys 159:290–311
Monaghan JJ (2005) Smoothed particle hydrodynamics. Rep Prog Phys 68:1703–1759
Monasse B (1995) Nucleation and anisotropic crystalline growth of polyethylene under shear. J Mater Sci 30:5002–5012
Mondy LA, Brenner H, Altobelli SA, Abbott JR, Graham AL (1994) Shear-induced particle migration in suspensions of rods. J Rheol 38:444–452
Moonan WK, Tschoegl NW (1985) The effect of pressure on the mechanical properties of polymers, IV. Measurements in torsion. J Polym Sci: Polym Phys Edn 23:623–651
Mori T, Tanaka K (1973) Average stress in matrix and average elastic energy of materials with misfitting inclusions. Acta Metall 21:571–574
Mori T, Wakashima K (1989) Successive iteration method in the evaluation of average fields in elastically inhomogeneous materials. In: Weng GJ, Taya M, Abe H (eds) Micromechanics and Inhomogeneity. Springer-Verlag, pp 269–282
Morland LW, Lee EH (1960) Stress analysis for linear viscoelastic materials with temperature variation. Trans Soc Rheol 4:233–263
Morris JP, Fox PJ, Zhu Y (1997) Modeling low Reynolds number incompressible flows using SPH. J Comput Phycs 136:214–226
Mura T (1987) Micromechanics of Defects in solids. Martinus Nijhoff Publishers, Boston
Mykhaylyk OO, Chambon P, Graham RS, Patrick J, Fairclough A, Olmsted PD, Ryan AJ (2008) The specific work of flow as a criterion for orientation in polymer crystallization. Macromol 41:1901–1904
Nakano R (1998) Apparatus and method for analyzing a process of fluid flow, an apparatus and method for analyzing an injection molding process, an injection molded product, and a production method of the injection molded product. US Patent No. 5835379
Nakano R (2000) Apparatus for analyzing a process of fluid flow, and a production method of an injection molded product. US Patent No. 6161057
Oldroyd JG (1950) On the formulation of rheological equations of state. Proc Roy Soc Lond A200:345–541
Osher S, Fedkiw R (2002) Level set methods and dynamic implicit surfaces. Springer-Verlag, New York
Osher S, Sethian JA (1988) Fronts propagating with curvature-dependent speed: Algorithms based on Hamilton-Jacobi formulations. J Comput Phys 79:12–49
Öttinger HC (1996) Stochastic processes in polymeric fluids, tools and examples for developing simulation algorithms. Springer, Berlin
Pantani R, Coccorullo I, Speranza V, Titomanlio G (2005) Modeling or morphology evolution in the injection molding process of thermoplastic polymers. Prog Polym Sci 30:1185-1122
Pantani R, Speranza V, Titomanlio G (2001a) Relevance of mold-induced thermal boundary conditions and cavity deformation in the simulation of injection molding. Polym Eng Sci 47:2022–2035
Pantani R, Speranza V, Titomanlio G (2001b) Relevance of crystallization kinetics in the simulation of injection molding process. Int Polym Process 16:61–71
Park SJ, Kwon TH (1996) Sensitivity analysis formulation for three-dimensional conduction heat transfer with complex geometrics using a boundary element method. Int J Numer Methods Eng 39:2837–2862

Park SJ, Kwon TH (1998) Thermal and design Sensitivity analysis cooling system of injection mold, Part 1: thermal analysis. J Manufact Sci Eng 120:287–295

Patankar SV (1980) Numerical heat transfer and fluid flow. McGraw-Hill, New York

Pearson JRA (1966) Mechanical Principles of Polymer Melt Processing. Pergamon Press, Oxford

Pearson JRA, Petrie CJS (1968) On the melt flow instability of extruded polymers. In: Wetton RE, Whorlow RW (eds) Polymer systems: deformation and Flow. Macmillan, London, pp 163–187

Pearson JRA, Richardson SM (eds) (1983) Computational analysis of polymer processing. applied science Publications, London

Phan-Thien N (1988) Influence of wall slip on extrudate swell: a boundary element investigation. J Non-Newtonian Fluid Mech 26:327–340

Phan-Thien N, Fan XJ (2002) Viscoelastic mobility problem using a boundary-element method. J Non-Newtonian Fluid Mech 105:131–152

Phan-Thien N, Fan XJ, Tanner RI, Zheng R (2002) Folgar–Tucker constant for a fiber suspension in a Newtonian fluid. J Non-Newtonian Fluid Mech 103:251–260

Phan-Thien N, Fan XJ, Zheng R (2000) A numerical simulation of suspension flow using a constitutive model based on anisotropic interparticle interactions. Rheol Acta 39:122–130

Phan-Thien N, Graham AL (1991) A new constitutive model for fiber suspensions: flow past a sphere. Rheol Acta 30:44–57

Phan-Thien N, Pham DC (2000) Differential multiphase models for polydispersed spheroidal inclusions: thermal conductivity and effective viscosity. Int J Eng. Sci 38:73–88

Phan-Thien N, Tanner RI (1977) A new constitutive equation derived from network theory. J Non-Newtonian Fluid Mech 2:353–365

Phan-Thien N, Zheng R (1997) Macroscopic modeling of evolution of fiber orientation during flow. In: Papathanasiou TD, Guell DC (eds) Flow-Induced Alignment in Composite Materials. Woodhead publishing Ltd, Cambridge, pp 77–111

Phelps JH, Tucker CL III (2009) An anisotropic rotary diffusion model for fiber orientation in short–and long-fiber thermoplastics. J Non-Newtonian Fluid Mech 156:165–176

Phillips RJ, Armstrong RC, Brown RA, Graham AL, Abbott JR (1992) A constitutive equation for concentrated suspension that accounts for shear-induced particle migration. Phys Fluids A2:30–40

Pichelin E, Coupez T (1998) Finite element solution of the 3D mold filling problem for viscous incompressible fluid. Comput Methods Appl Mech Eng 163:359–371

Prakash M, Cleary P, Grandfield J (2009) Modeling of metal flow and oxidation during furnace emptying using smoothed particle hydrodynamics. J Mater Proc Tech 209:3396–3407

Qiao H (2005) Transient mold cooling analysis using BEM with the time-dependent fundamental solution. Int Commun Heat and Mass Trans 32:315–322

Rafiee A (2008) Modeling of generalized Newtonian lid-driven cavity flow using an SPH method. ANZIAN J 49:411–422

Rafiee A, Manzari MT, Hosseini M (2007) An incompressible SPH method for simulation of unsteady viscoelastic free-surface flows. Int J Non-linear Mech 42:1210–1223

Rajupalem V, Talwar K, Friedl C (1997) Three-dimensional simulation of the injection molding process. Society of Plastics Engineers Annual Technical Conference—ANTEC'1997 43:670–673

Ramamurthy AV (1986) Wall slip in viscous fluids and influence of materials of construction. J Rheol 30:337–357

Ranganathan S, Advani SG (1991) Fiber-fiber interactions in homogeneous flows of nondilute suspensions. J Rheol 35:1499–1522

Reiner M (1964) Deborah number. Phys Today 17:62

Rezayat M, Burton T (1990) Boundary-integral formulation for complex three-dimensional geometries. Int J Numer Methods Eng 29:263–273

Rice JG, Schnipke RJ (1986) An equal-order velocity-pressure formulation that does not exhibit spurious pressure modes. Comput Methods Appl Mech Eng 58:135–149

Richardson SM (1972) Hele Shaw flows with a free boundary produced by the injection of fluid into a narrow channel. J Fluid Mech 56:609–618

Rizzo FJ, Shippy DJ (1970) A method of solution for certain problems of transient heat conduction. AIAA J 8:2004–2009

Roscoe R (1952) The viscosity of suspensions of rigid spheres. Br J Appl Phys 3:267–269

Rosen BW, Hashin Z (1970) Effective thermal expansion coefficients and specific heats of composite materials. Int J Eng Sci 8:157–173

Rosenbaum EE, Hatzikiriakos SG (1997) Wall slip in the capillary flow of molten polymers subject to viscous heating. AIChE J 43:598–608

Rubin II (1972) Injection molding–theory and practice. Wiley, New York

Santhanam N, Chiang HH, Himasekhar K, Tuschak P, Wang KK (1991) Postmolding and load-induced deformation analysis of plastic parts in the injection molding process. Adv Polym Tech 11:77–89

Sato T, Richardson SM (1995) Numerical simulation of the fountain flow problem for viscoelastic fluids. Polym Eng Sci 39:805–812

Schieber JD, Venerus DC, Bush K, Balasubramanian V, Smoukov S (2004) Measurement of anisotropic energy transport in flowing polymers using a holographic technique. Proc Nat Acad Sci 101:13142–13146

Sethian JA (1999) Level set methods and fast marching methods: evolving interfaces in computational geometry, fluid mechanics, computer vision and materials science. Cambridge University Press, Cambridge

Sethian JA, Smereka P (2003) Level set methods for fluid interfaces. Annu Rev Fluid Mech 35:341–372

Shao S, Lo EYM (2003) Incompressible SPH method for simulating Newtonian and non-Newtonian flows with a free surface. Advan Water Resources 26:787–800

Sherbelis G, Friedl C (996) The importance of pressure dependent viscosity and contraction pressure losses in injection molding CAE analysis. Society of Plastics Engineers Annual Technical Conference—ANTEC'1996 42:778–782

Shoemaker J (2006) Moldflow design guide: a reference for plastics engineers. Hanser, Munich

Sidi S, Kamal MR (1982) Thermoset injection molding dynamics and the distribution of cure in thermoset molded parts. Polym Eng Sci 22:349–353

Singh P, Kamal MR (1989) The effect of processing variables on microstructure of injection molded short fiber reinforced polypropylene composites. Polym Compos 70:344–351

Speight RG, Costa, FS, Hadinata C (2010) Pressure and temperature dependence of thermal conductivity for injection molding simulation. Proceedings of the 26th annual conference of the polymer processing society, Banff, Canada

Struik LCE (1990) Internal stresses, dimensional instabilities and molecular orientations in plastics. Wiley, Chichester

Sun J, Phan-Thien N, Tanner RI (1996) An adaptive viscoelastic stress splitting scheme and its applications: AVSS/SI and AVSS/SUPG. J Non-Newtonian Fluid Mech 65:75–91

Sun J, Smith MD, Armstrong RC, Brown RA (1998) Finite element method for viscoelastic flows based on the discrete adaptive viscoelastic stress splitting and the discontinuous Galerkin method: DAVSS-G/DG. J Non-Newtonian Fluid Mech 86:281–307

Sussman M, Fatemi E, Smereka P, Osher S (1998) An improved level set method of incompressible two-phase flows. Comput Fluids 27:663–680

Sussman M, Smereka P, Osher S (1994) A level set approach to computing solutions to incompressible two-phase flow. J Comput Phys 114:146–159

Tadmor Z, Broyer E, Gutfinger C (1974) Flow Analysis Network (FAN)-Solving Flow Problems in Polymer Processing. Polym Eng Sci 14:660–665

Talwar K, Costa FS, Rajupalem V, Antanovski LK, Friedl C (1998) Three dimensional simulation of plastic injection molding. Society of Plastics Engineers Annual Technical Conference—ANTEC'1998 44:562–566

Tandon GP, Weng GJ (1984) The effect of aspect ratio of inclusion on elastic properties of unidirectionally aligned composites. Polym Compos 5:327–333

Tanner RI (1960) Full-film lubrication theory for a Maxwell liquid. Int J Mech Sci 1:206–215

Tanner RI (1975) Stresses in dilute solutions of bead-nonlinear-spring macromolecules. ii. Unsteady flows and approximate constitutive relations, constitutive relations. J Rheol 19:37–65

Tanner RI (1982) The stability of some numerical schemes for model viscoelastic fluids. J Non-Newtonian Fluid Mech 10:169–174

Tanner RI (1994) Partial wall slip in polymer flow. Ind Eng Chem Res 33:2434–2436

Tanner RI (2000) Engineering rheology, 2nd edn. Oxford University Press, New York

Tanner RI (2003) On the flow of crystallizing polymers, I. Linear regime. J Non-Newtonian Fluid Mech 112:251–268

Tanner RI (2006) On the congruence of some network and pom-pom models. Korea-Australia Rheol J 18:9–14

Tanner RI (2009) The changing face of rheology. J Non-Newtonian Fluid Mech 157:131–141

Tanner RI, Nasseri S (2003) Simple constitutive models for linear and branched polymers. J Non-Newtonian Fluid Mech 116:1–17

Tanner RI, Qi F (2005a) A comparison of some models for describing polymer crystallization at low deformation rate. J Non-Newtonian Fluid Mech 127:131–141

Tanner RI, Qi F (2005b) A phenomenological approach to suspensions with viscoelastic matrices. Korea-Australia Rheol J 17:149–156

Tanner RI, Qi F (2009) Stretching, shearing and solidification. Chem Eng Sci 64:4576–4579

Tanner RI, Qi F, Housiadas KD (2010a) A differential model for the rheological properties of concentrated suspensions with weakly viscoelastic matrices. Rheol Acta 49:169–176

Tanner RI, Qi F, Housiadas KD (2010b) A differential approach to suspensions with power-law matrices. J Non-Newtonian Fluid Mech 165:1677–1681

Tanner RI, Walters K (1998) Rheology: An historical perspective. Elsevier, Amsterdam

Tanner RI, Xue SC, Phan-Thien N (1995) Macromolecules and megabytes. In: Shen SF, Dawson PR (eds) Simulation of materials processing: theory, methods and applications. NUMIFORM 95. Balkema, Rotterdam, pp 165–172

Titomanlio G, Drucato V, Kamal MR (1987) Mechanism of cooling stress build-up in injection molding of thermoplastic polymers. Int Polym Process 1:55–59

Toor HL, Ballman RL, Copper L (1960) Predicting mold flow by electronic computer. Modern Plastic 38:117–120

Tucker CL III (1991) Flow regimes for fiber suspensions in narrow gaps. J Non-Newtonian Fluid Mech 39:239–268

Tucker CL III, Advani SG (1994) Processing of fiber systems. In: Advani SG (ed) Flow and rheology in polymer composites manufacturing. Elsevier, Amsterdam, pp 147–202

Tucker CL III, Liang E (1999) Stiffness predictions for unidirectional short-fiber composites: Review and evaluation. Comp Sci Technol 59:655–671

Tucker III CL, Wang J, O'Gara JF (2007) Method and article of manufacture for determining a rate of change of orientation of a plurality of fibers disposed in a fluid. US Patent 7266469

Van den Brule BHAA (1989) A network theory for the thermal conductivity of an amorphous polymeric material. Rheol Acta 28:257–266

Van den Brule BHAA (1990) The non-isothermal elastic dumbbell: a model for the thermal conductivity. Rheol Acta 29:416–422

Van den Brule BHAA, O'Brien SBG (1990) Anisotropic conduction of heat in a flowing polymeric material. Rheol Acta 29:580–587

Van der Beek MHE, Peters GWM, Meijer HEH (2006) Influence of shear flow on the specific volume and the crystalline, morphology of isotactic polypropylene. Macromol 39:1805–1814

Van Krevelen DW (1976) Properties of polymers: their estimation and correlation with chemical structure, 2nd edn. Elsevier, Amsterdam

Van Leer B (1979) Towards the ultimate conservative difference scheme. J Comput Phys 32:101–136

Van Meerveld J, Peters GWM, Hutter M (2004) Towards a rheological classification of flow induced crystallization experiments of polymer melts. Rheol Acta 44:119–134

Venerus DC, Schieber JD, Balasubramanian V, Bush K, Smoukov S (2004) Anisotropic thermal conduction in a polymer liquid subjected to shear flow. Phys Rev Lett 93:098301

Venerus DC, Schieber JD, Iddir H, Guzman JD, Broerman AW (1999) Relaxation of anisotropic thermal diffusivity in a polymer melt following step shear strain. Phys Rev Lett 82:366–369

Venerus DC, Schieber JD, Iddir H, Guzman JD, Broerman AW (2000) Evidence of a stress-thermal rule in molten poly-isobutylene following a step shearing strain. AIChE J 46:610–615

Venerus DC, Schieber JD, Iddir H, Guzman JD, Broerman AW (2001) Anisotropic thermal diffusivity measurements in deforming polymers and the stress-thermal rule. Int J Thermophys 22:1215–1225

Verbeeten WMH, Peters GWM, Baaijens FPT (2001) Differential constitutive equations for polymer melts: The extended Pom-Pom model. J Rheol 45:823–843

Verbeeten WMH, Peters GWM, Baaijens FPT (2002) Viscoelastic analysis of complex polymer melt flows using the extended Pom-Pom model. J Non-Newtonian Fluid Mech 108:301–326

Verleye V, Dupret F (1993) Prediction of fiber orientation in complex injection molded parts. Proc ASME Winter Annual Mtg, New Orleans 175:139–163

VerWeyst BE (1998) Numerical Predictions of Flow-induced Fiber Orientation in 3-D Geometries. Ph.D Thesis, University of Illinois at Urbana-Champaign, Urbana, IL

VerWeyst BE, Tucker CL III, Foss PH, O'Gara JF (1999) Fiber orientation in 3-D injection molded features: prediction and experiment. Int Polym Process 14:409–420

Walsh SF (1993) Shrinkage and warpage prediction for injection molded components. J Reinforced Plast Compos 12:769–777

Wang J, O'Gara JF, Tucker CL III (2008) An objective model for slowing orientation kinetics in concentrated fiber suspensions: theory and rheological evidence. J Rheol 52:1179–1200

Wang VW, Hieber CA, Wang KK (1986) Dynamic simulation and graphics for the injection molding of 3-dimensional thin parts. J Polym Eng 7:21–45

Warner HR Jr (1972) Kinetic Theory and rheology of dilute suspensions of finitely extendible dumbbells. Ind Eng Chem Fundam 11:379–387

Wassner E, Maier RD (2000) Shear-induced crystallization of polypropylene melts. Proc XIII Int Congr Rheol, Cambridge. 1:183–185

White JL (1964) Dynamics of viscoelastic fluids, melt fracture, and the rheology of fiber spinning. J Appl Polym Sci 8:2339–2357

Whiteford B (1992) Success through Excellence, Chapter 35: Moldflow. Beunont Publishing House, South Perth, pp 387–396

Williams G, Lord HA (1975) Mold-filling studies for the injection molding of thermoplastic materials. Part I: The flow of plastic materials in hot- and cold-walled circular channels. Polym Eng Sci 15:553–568

Williams ML, Landel RF, Ferry JD (1955) The temperature dependence of relaxation mechanisms in amorphous polymers and other glass-forming liquids. J Amer Chem Soc 77:3701–3707

Yu H, Kennedy PK (2004) Apparatus and methods for performing process simulation using a hybrid model. US Patent Application, No. 20040210429

Yu H, Kietzmann C, Cook P, Xu S, Costa FS, Kennedy PK (2004) A new method for simulation of injection molding. Society of Plastics Engineers Annual Technical Conference—ANTEC' 2004(50):641–644

Yu H, Thomas R (2000) Method for modeling three-dimensional objects and simulation of fluid flow. United States Patent No. 6096088

Zarraga IE, Hill DA, Leighton D (2001) Normal stress and free surface deformation in concentrated suspensions of noncolloidal spheres in a viscoelastic fluid. J Rheol 45:1065–1084

Zhai M, Lam YC, Au CK (2006) Runner sizing and weld line positioning for plastics injection molding with multiple gates. Eng Comput 21:218–224

Zhang HL, Ong NS, Lam YC (2008) Experimental investigation of key parameters on the effects of cavity surface roughness in microinjection molding. Polym Eng Sci 48:490–495

Zhao HK, Chan T, Merriman B, Osher S (1996) A variational level-set approach to multiphase motion. J Comput Phys 127:179–195

Zheng R (1991) Boundary Element Method for Some Problems in Fluid Mechanics and Rheology. Ph.D Thesis, The University of Sydney

Zheng R, Coleman CJ, Phan-Thien N (1991a) A boundary element approach for non-homogeneous potential problems. Comput Mech 7:279–288

Zheng R, Hadinata C, Kennedy PK (2008) Numerical study on the impact of additives on shrinkage of injection molded polypropylene. The XV Int Congr Rheol, Monterey, AIP Conf Proc 1027:45–47

Zheng R, Kennedy PK (2001) Numerical Simulation of Crystallization in Injection Molding. Proc. 6th World Congr Chem Eng, Pap110. Melbourne

Zheng R, Kennedy PK (2004) A model for post-flow induced crystallization: general equation and predictions. J Rheol 48:823–842

Zheng R, Kennedy PK (2006) Anisotropic thermal conduction in injection molding. Proc. 22nd Annual Conf Polym Process Soc, Yamagata

Zheng R, Kennedy PK, Fan XJ, Phan-Thien N, Tanner RI (2000) Brownian dynamics simulation of non-dilute fiber suspension flows in a centre-gated disk. Society of Plastics Engineers Annual Technical Conference—ANTEC'2000, Orlando, Florida, 46: 671-674

Zheng R, Kennedy PK, Phan-Thien N, Fan XJ (1999) Thermoviscoelastic simulation of thermally and pressure induced stresses in injection molding for the prediction of shrinkage and warpage for fiber-reinforced thermoplastics. J Non-Newtonian Fluid Mech 84:159–190

Zheng R, McCafferey N, Winch K, Yu H, Kennedy PK (1996) Predicting warpage of injection molded fiber-reinforced plastics. J Thermoplastic Compos Mater 9:90–106

Zheng R, Phan-Thien N (1992) Transforming domain integrals to the boundary using approximate particular solutions. Appl Numer Math 10:435–445

Zheng R, Phan-Thien N, Coleman CJ (1991b) A boundary element approach for non-linear boundary-value problems. Comput Mech 8:71–86

Zheng R, Phan-Thien N, Tanner RI (1990) The flow of a suspension fluid in injection molding. In: Yeow YL, Uhlherr PHT (ed) Proc Fifth Natl Conf Soc Rheol , Melbourne, pp 141–144

Zheng R, Phan-Thien N, Tanner RI, Coleman CJ (1992) A boundary element/particular solution approach for non-Newtonian flow problems. In: Moldenaers P, Keunings R (eds) Theoretical and applied rheology. Proceedings of the XI international congress on rheology, Brussels, Belgium, p 311

Zheng R, Tanner RI, Lee Wo D, Fan XJ, Hadinata C, Costa FS, Kennedy PK, Zhu P, Edward G (2010) Modeling of flow-induced crystallization of colored polypropylene in injection molding. Korea-Australia Rheol J 22:151–162

Zhou H, Li DQ (2004) Modeling and prediction of weld line location and properties based on injection molding simulation. Int J Mater Prod Technol 21:526–538

Zhou H, Zhang Y, Wen J, Li DQ (2009) An acceleration method for the BEM-based cooling simulation of injection molding. Engineering Analysis with Boundary Elements 33:1022–1030

Zhu H, Martys NS, Ferraris C, Kee DD (2010) A numerical study of the flow of Bingham-like fluids in two-dimensional vane and cylinders using a smoothed particle dynamics (SPH) based method. J Non-Newtonian Fluid Mech 165:362–375

Zhu P, Phillips A, Edward G, Nichols L (2009) Experimental observation of effects of seeds on polymer crystallization. Phys Rev E 80:051801

Ziabicki A (1996) Crystallization of polymers in variable external conditions, II, Effect of cooling in the absence of stress and orientation. Colloid Polym Sci 274:705–716

Zienkiewicz OC, Taylor RL, Zhu JZ (2005) The Finite Element Method: Its Basis and Fundamentals, 6th edn. Elsevier Butterworth-Heinemann, Burlington

Zoetelief WF, Douven LFA, Ingen-Housz AJ (1996) Residual thermal stresses in injection molded products. Polym Eng Sci 36:1886–1896

Zoller P, Fakhreddine YA (1994) Pressure-volume-temperature studies of semi-crystalline polymers. Thermochim Acta 238:397–415

Zoller P, Kehl TA, Starkweather HW, Jones GA (1989) The equation of state and heat of fusion of poly(ether ether ketone). J Polym Sci Part B: Polym Phys 27:993–1007

Zuidema H (2000) Flow Induced Crystallization of Polymers: Application to Injection Molding. Ph.D Thesis, Eindhoven University of Technology

Zuidema H, Peters GWM, Meijer HEH (2001) Development and validation of a recoverable strain based model for flow-induced crystallization of polymers. Macromol Theory Simul 10:447–460

Author Index

R. Zheng et al., *Injection Molding*, DOI: 10.1007/978-3-642-21263-5,

Subject Index

MIX
Papier aus verantwortungsvollen Quellen
Paper from responsible sources
FSC® C105338

If you have any concerns about our products,
you can contact us on
ProductSafety@springernature.com

In case Publisher is established outside the EU,
the EU authorized representative is:
Springer Nature Customer Service Center GmbH
Europaplatz 3, 69115 Heidelberg, Germany

Printed by Libri Plureos GmbH
in Hamburg, Germany